# HISTORY

### of the

## 45th: 1st NOTTINGHAMSHIRE REGIMENT

### (SHERWOOD FORESTERS)

BY

## Col. P. H. DALBIAC

### Late Captain 45th Regiment

SWAN SONNENSCHEIN & CO., LTD.

PATERNOSTER SQUARE, E.C.

1902

# PREFACE

THE larger share of the credit for the work in the following pages is due to the late Admiral Colomb, who was asked some twelve years ago by Colonel Hooke, who was then commanding the 1st battalion Sherwood Foresters, formerly the 45th regiment, to undertake the task. The undertaking was by no means a light one; Admiral Colomb was practically asked to make bricks without straw, as hardly any material was forthcoming. But he manfully faced the work, and at the time of his death had collected a large part of the material necessary, and constructed the history for a period of nearly one hundred years.

When I was asked some years later by Colonel Hume, who had succeeded to the command of the regiment, to complete the work for press, I found the task as light as it was congenial. The admirable work done by Admiral Colomb was handed over to me, and I had only to select from the material collected by him such facts as I considered essential to a true and succinct account of the doings of the regiment, and after verification to arrange them in the form in which they now appear.

For the period subsequent to 1859, for which I could obtain no records or material of any sort, I am largely indebted to the kind assistance of officers of the regiment who were serving during that period, and should like to express my especial obligation, among others, to Major-General Hayward, Captain Reeve-King, Captain Patterson, and the late Major Higgins, without whose friendly help I should indeed have been in a difficult position. I have also been much indebted to Mr. Wilfred Brinton for his kind loan of a book entitled, " The British Army : As it is and as it ought to be," by Major Campbell, an adjutant of the 45th regiment during the Peninsular War, from which I have been able to collect many interesting incidents affecting the regiment.

The illustration of the monument to the regiment at Secunderabad is from a sketch by the late Major Baldwin, R.E., while Mr. Frank Wallis has been responsible for the pictures illustrating the uniform of the regiment at different periods.

It is in the hope that the records of untarnished honour and uniform soldierlike and good behaviour, in peace as well as on the battlefield, which these pages contain, may arouse feelings of pride in those who have borne their share in them in the past, and call forth a spirit of emulation in those who, as the successors of the old 45th regiment, have now in their hands the power to add fresh lustre to the laurels gained by their predecessors—a hope which, I am bold enough to say, has been amply fulfilled in South Africa —I dedicate this short history to the memory of sixteen years spent in the old regiment, with a confident trust in the kindly criticism of all into whose hands the book may fall.

                                    P. H. DALBIAC.

MAY, 1902.

# CONTENTS

# CONTENTS

# HISTORY OF THE 45TH REGIMENT

## CHAPTER I

1739 to 1747—The Marine Battalions—Hannah Snell—
The 56th become the 45th.

THE title of "45th Regiment" was first borne by the second of ten "marine regiments" raised on the outbreak of war with Spain in 1739. At the close of the war, nine years later, these regiments were disbanded; consequently the "45th Regiment" of later years had no connection with this regiment; nevertheless, it seems hardly out of place, in a history of the regiment which bore the same number for nearly a century and a half, to give a slight sketch of the doings of the original Forty-fifth.

The old 45th regiment was commonly called "The Green Marines," from their green facings and green, high-crowned leather caps; and, when first raised, consisted nominally of 1000 men, and was commanded by Colonel Robinson. The other nine battalions raised at the same time and soon after, were of the same strength, and were commanded by Colonels Wolfe, Lowther, Wynyard, Douglas, Moreton, Pawlet, Cornwall, Hanmere, and Jeffreys. Six, at least, of the regiments were included in the force under Lord Cathcart which sailed in October, 1740, with the fleet under Sir Chaloner Ogle, to reinforce Admiral Vernon in the West Indies.

The fleet arrived at Dominica on the 19th December, and about the same time an almost irreparable loss was sustained

A

in the death of Lord Cathcart; the command of the land and
marine forces devolving on Brigadier-General Wentworth.

In March, 1741, the whole force proceeded to the attack
of the strongly-fortified town of Carthagena, and which ended
in disaster; 5000 troops were landed for the assault on the
10th March, but only 3200 were re-embarked a month later,
on the 16th April, when the attempt on the place was
abandoned.

The force, or what remained of it, was subsequently landed
in July, 1741, at Port Cumberland, in St. Jago de Cuba, on
the south end of the island, with the intention of conquering
it.   Differences, however, unfortunately arose between the
Admiral and Wentworth; the troops remained inactive, and
were re-embarked on the 20th of November without having
fired a shot.

Colonel Robinson having been killed in a repulse from the
Castle of St. Lagore, the key to the approach to Carthagena,
Lieut.-Colonel R. Frazer succeeded to the command of the
regiment, which, on the evacuation of Cuba, had dwindled
down to a strength of only 158 of all ranks, excepting officers,
namely, 22 sergeants, 21 corporals, 6 drummers, and 109
privates.

So great had been the losses of all the regiments through
sickness that, although on the 15th January, 1742, a reinforcement of 2000 men reached Wentworth, he declared
on the 22nd that he had not more than 3000 men fit for duty.

In March, however, fleet and troops proceeded to attack
Panama, through Portobello, and on the 28th reached the
latter place unopposed.   Here differences of opinion
between the Admiral and Wentworth again broke out, with
the result that the whole expedition was brought back, and
reached Jamaica in the middle of May.   Colonel Frazer had
taken part in the council of war which decided on the 30th
March that the attack on Panama was impracticable.

· Before August, the relations between Vernon and Wentworth had become quite hopeless, and on September 23rd, 1742, orders — unhappily welcome to both officers — were received directing their return to England. Things had got to such a pass between them that the Admiral told Wentworth "that to his inexperience, want of judgment, and unsteady temper was principally owing His Majesty's affairs having prospered so ill in those parts."

On Vernon and Wentworth quitting the West Indies a certain portion of the marines were placed in the fleet directly under the orders of Sir Chaloner Ogle. They were subsequently employed with Dalzell's Regiment, under Commodore Knowles, in the disastrous attack on Port Cavallo. A portion of the marines were afterwards engaged at the siege of Louisbourg and the conquest of Cape Breton in May, 1745.

At the close of the year 1746, the whole of the marine regiments appear to have been in Great Britain, but, in 1747, they were apparently employed in India at the siege of Pondicherry.

A story is told of a woman, named Hannah Snell, having enlisted in the regiment at Gosport. She had, it is said, previously enlisted in, and deserted from, a regiment in the North of England. She embarked on board the "Swallow," one of the fleet with which Boscawen sailed to the East Indies, and behaved with distinguished courage on several occasions, receiving a ball in the groin, which she extracted herself two days afterwards. Eleven other wounds necessitated her removal first to hospital at Cuddalore and then to England, in spite of which she managed to keep her secret until she was discharged. Subsequently she revealed it, and in consequence of a petition to H.R.H. the Duke of Cumberland, she was granted a pension of thirty pounds a year for life.

At the peace of Aix-la-Chapelle the marine regiments were

disbanded, and seven new line regiments, which had been raised in 1741, and numbered from 54 to 60, were moved up into their places.     Thus the 56th regiment became the 45th of the line, and its history has since that time been continuous as the 45th, 1st Nottinghamshire Regiment, the Sherwood Foresters, down to 1881, when it became, under Mr. Childers' new territorial scheme, the First Battalion of the Derbyshire Regiment.

# CHAPTER II

THE 56th regiment, which afterwards became the 45th, raised in the year 1741, probably wore the usual infantry uniform of the day: a square-cut red coat, laced down the front, with broad facings, which in the case of this regiment were dark green; the ample skirts were looped back to show the lining and waistcoat; white breeches and gaiters, buff leather belts and accoutrements, with three-cornered hats. The clothing was richly decorated with parti-coloured lace, the same as is still worn at the present day by drummers. The Grenadier companies wore the Prussian sugar-loaf cap, while the light companies probably wore the bugle, which had been adopted as their distinguishing badge as early as 1656.

The arms were a flint-lock musket, bayonet, and sword. The Grenadiers no longer used the grenade, but the officers still carried the spontoon, a light partisan, having an axe on one side and a pick on the other, the weapon now carried by the Yeomen of the Guard.

Captain Daniel Houghton, of the 1st Foot Guards, was appointed to the command of the regiment on its formation, and consequently it was usually known, after the custom of those days, as " Houghton's Regiment."

One of the earliest duties of the regiment was to send a detachment under a lieutenant, in October, 1741, to Tilbury Fort, as a guard to the recruits of Dalzell's Regiment.

In December of the same year the headquarters with five companies were at Canterbury, and detachments were at Wye and Ashford, which, early in the following year, were withdrawn, and others sent to Hastings, Sandwich, and Rye.

At the end of April the regiment, now ten companies strong, received orders to embark at Portsmouth, and was broken up into detachments at Chichester, Portsmouth, Fareham, Waltham, Havant, Alton, Arundel, and Midhurst, awaiting transport. Unfortunately, their destination remains a mystery, as no trace can be found of whither it was intended that the regiment should proceed from Portsmouth, though the fact that "Houghton's Regiment" was at Gibraltar in 1745 may indicate that that place was its destination. On the other hand, when Houghton was succeeded in the command of the 56th by Warburton in June, 1745, he was appointed "to be Colonel of the Regiment of Foot, late under the command of Thomas Wentworth." Whether it was this regiment or the 56th regiment which was at Gibraltar there seems.to be no definite means of deciding.

In the autumn of 1745 the headquarters of "Warburton's Regiment" were at Plymouth, with detachments at Bristol and Swansea. In the same year we find the Bristol detachment called upon to furnish an escort for certain treasure which had arrived from Kinsale, en route for London, while the Swansea detachment was employed in receiving pressed men from the county of Glamorgan. It rejoined headquarters in September.

In this year the carrying of a sword was discontinued, except by Grenadiers.

There is a tradition that the regiment was at the battle of Prestonpans in September of this year; and it is more definitely stated that five companies were in Scotland early in 1746, at Dumfries, Stranraer, Stirling, and Glasgow. This had probably arisen through a confusion with "Houghton's

Corps," which was ordered to embark at Tilbury, for North Britain, on the 27th of March, 1746, as it is practically certain that the 56th, or "Warburton's Regiment," embarked for Cape Breton on the 28th of October, 1745, being 655 strong.

In 1746 soldiers ceased to wear wigs, and those whose hair was long were ordered to tuck it up under their hats. Officers were ordered to mount guard in queue wigs, or with their hair tied. Brown cloth gaiters were worn, and a mixture of yellow ochre and whiting was used to colour the belts.

In 1748, as has been mentioned before, owing to the disbandment of the marine battalions, the 56th regiment—still commanded by Warburton—became the 45th, and two years later we find it quartered in Nova Scotia. In this year Major Lawrence was promoted to the Lieut.-Colonelcy of "Cornwallis' Regiment," and made Governor of Annapolis Royal.

In July, 1751, a Royal Warrant was issued confirming the numerical titles of regiments, and thus in a great measure abolishing the somewhat confusing practice of designating them by the names of their colonels.

In the same warrant it was directed that the numbers were to be borne on the centre of the colours, surmounted by a Royal Crown. Some years passed before the numbers appeared on the accoutrements, but the buttons of this year are stated to have had the regimental number embossed upon them for the first time. The first complete and authentic list of officers of the regiment dates from 1754, in which year also began the issue of the regular annual army lists.

Ever since the peace of Aix-la-Chapelle, France had been pursuing in North America what was considered by Great Britain to be a hostile and unfair policy. Territories with doubtful claims to them were seized, and difficulties thrown in the way of British trade; disputes about boundaries arose, and M. de la Galissonière, in the assertion of what he considered the rights of France, erected a fort at Beau Sejour, at the head of the Bay of Fundy.

In June, 1755, it was determined to oust the French from their position. Accordingly, an expedition consisting of some 2000 men, under Lieut.-Colonel Monckton, escorted by some frigates, moved up the bay and attacked and captured the fort. The 45th, or part of it, was employed with this force, and probably for the first time took part in active operations in the field. Colonel Monckton, after putting a garrison into the place, and changing its name to Cumberland, proceeded the next day to the attack of another French fort upon the river Gasperan, which runs into Bay Verte.

This fort was the chief depôt for supplying the French Indians and Arcadians with arms, ammunition, and other necessaries. Monckton speedily effected its capture, and all the stores fell into the hands of the English. The Arcadians, to the number of 15,000, were disarmed, and the expedition brought to a successful conclusion at the cost of only 20 killed and a like number wounded; while peace and tranquillity were restored and preserved to Nova Scotia.

The regiment continued to serve in the "Plantations," as the colonies were called, up to 1758, in which year it was engaged at the capture of Louisbourg by Lord Amherst.

At Louisbourg Lieut.-Colonel Wilmot was in command, and there were present besides:—1 major, 7 captains, 17 lieutenants, 6 ensigns, 1 adjutant, 1 quartermaster, 1 surgeon, 1 surgeon's mate, 38 sergeants, 19 drummers, and 852 rank and file. The expedition consisted in all of 13,225 men, of whom 571 were officers, and was under the command of Major-General Amherst, who had under him as brigadiers Colonels Whitmore, Lawrence, and Wolfe. The staff consisted of Lieut.-Colonel the Hon. Roger Townsend, Adjutant-General; Lieut.-Colonel Robertson, Quartermaster-General; Colonel

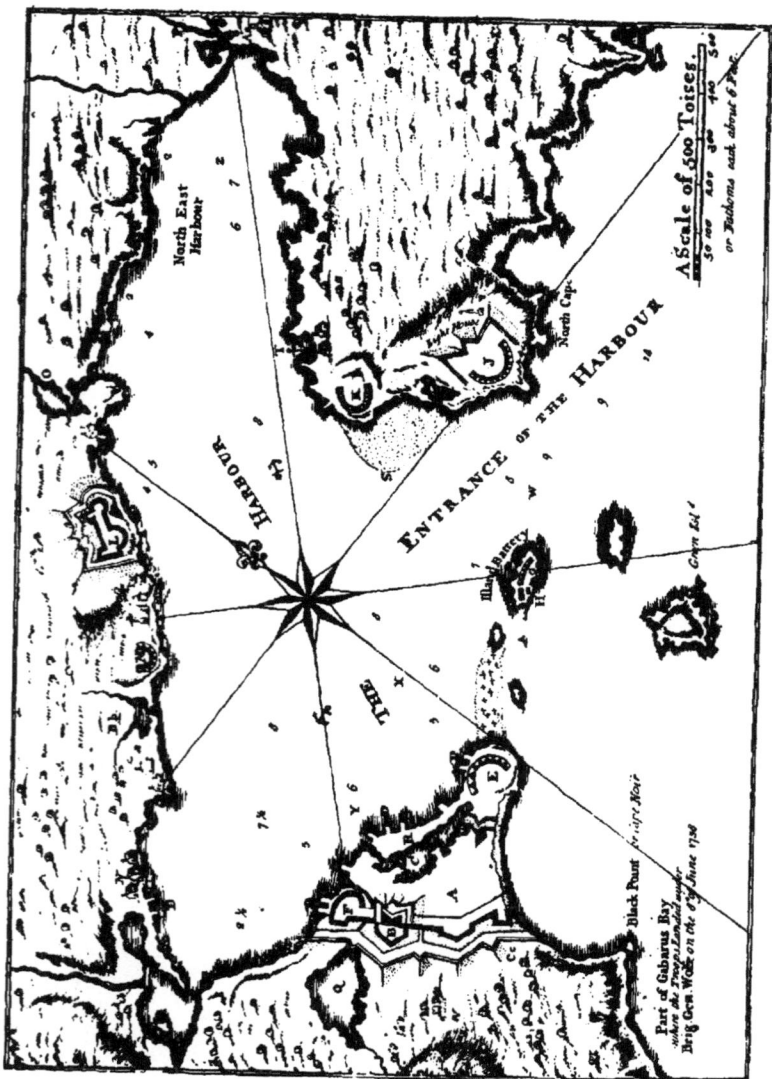

Williamson commanded the artillery and Colonel Bastide the engineers.

The troops were landed in Cabarus Bay, where the enemy had made every preparation to receive them, while their landing was much delayed by the heavy surf.

The army disembarked on the 8th of June in three columns, the 45th being in the centre column, which was commanded by Brigadier-General Lawrence. The landing was covered by the frigates, but the determined opposition of the enemy caused heavy loss to our troops. Eventually the French fell back, but fire was not opened on their works until the 21st of July, when five days' bombardment proved sufficient to compel the French commander, the Chevalier de Dricourt, to capitulate. Three thousand and thirty-one men, including 214 officers, with 443 sick and wounded, were made prisoners.

In 1759 the regiment remained stationed at Louisbourg, but two Grenadier companies formed part of the 3rd, or Murray's brigade, at the capture of Quebec. They were on the right, with "Bragg's Regiment," under the immediate command of Wolfe. These companies were directed to wear black feathers, presumably as mourning for General Wolfe, and possibly as distinction for their gallant conduct on the Heights of Abraham.

In 1761 the command of the regiment was twice changed, General Andrew Robinson appearing as Colonel in September, and General the Hon. John Boscawen in November.

In the following year a French squadron with a body of troops left Brest for the conquest of Newfoundland. Appearing off the coast at the head of 1500 men, de Haussonville took possession of the capital, St. John's, placed the fort in the best possible state of defence, and threw a boom across the narrow entrance to the harbour.

Commodore Lord Colville of Culross, who was at Halifax,

on being informed of these proceedings, lost no time in sailing to the relief of the island. He collected troops from New York, Halifax, and Louisbourg, a detachment of the 45th being included from the last-named place; the whole were placed under the command of Colonel William Amherst.

The force sailed for St. John's under convoy of six men-of-war, and was landed on the 13th of September at Torbay, seven miles to the northward of St. John's, while the ships blockaded the harbour. The French made an attempt at opposition, but Amherst drove them back, and, pushing on, took possession of a strong position opposite Signal Hill, where he placed a mortar battery. Some fighting followed, but de Haussonville soon capitulated; the enemy's fleet, unfortunately, escaping under cover of a fog. Our loss was about fifty killed and wounded, and our dead lie in the little enclosure on the northern slope of Signal Hill. Fort Amherst, on the south side of the narrows of St. John's, commemorates in its name this successful operation, in which the 45th took a prominent part.

On the 13th of August in this year the regiment appears in the returns of H.M. forces in America as "Boscawen's Regiment," with a strength of 771, or but 24 short of the establishment.

In 1763, after the signature of the Treaty of Peace at Paris, all regiments junior to the 70th were reduced.

The uniform of line regiments at this time was of a single type. Three-cornered cocked hats, bound with white lace and ornamented with a white loop and the black cockade of the House of Hanover; the scarlet coat was lined with the colour of the regimental facings—green, in the case of the 45th—and laced with white; the vests and breeches were scarlet; and long white gaiters were worn; the buttons were of flat metal, and without the embossed regimental number.

Lieutenant Alexander Ross, who joined the regiment in 1764, had entered the army in the 50th Foot in 1760, and taken part in all the actions of the allied army in Germany at the beginning of the war, and all the principal actions in America, where he had acted as A.D.C. to Lord Cornwallis. After the American war, while still on the strength of the regiment, he was for some time Deputy Adjutant-General in Scotland, and subsequently became Adjutant-General to the forces in the East Indies. In this year the swords, till then carried by the Grenadier companies, were abolished.

The regiment quitted America in 1765, and landed in Ireland on the 10th of July.

General William Haviland, who appears as Colonel of the regiment in 1767, had been engaged at the capture of Ile aux Noix and the reduction of Montreal in August and September, 1760, operations which completed the conquest of the great province of Canada.

In 1769 the uniform of the regiment appears in the army list as red, faced with deep green; white lace, with one green stripe.

The regiment appears to have remained in Ireland for more than ten years, but was in New York in 1776, for in July of that year we hear of it being there, and destined to join the army under Lord Howe's command at Staten Island.

In August the first brigade of this army under Major-General Pigott, consisting of the 4th, 15th, 27th, and 45th regiments, opened the campaign by land; and at Long Island and in the action near Brooklyn, the 45th formed part of the right wing under General Clinton, which bore the brunt of the battle.

In the same year the officers and sergeants of the infantry were ordered to discontinue the use of the spontoon and halbert, and to carry fusils instead. This alteration, however, was only temporary as far as the sergeants were concerned, for they resumed the more ancient weapon after the war.

In the following year half-gaiters of black cloth were substituted for the long white gaiters previously worn.

The 45th returned to England in 1778, its strength on landing being barely 100, and in the following year was quartered at Chatham.

A most important event in the history of the regiment occurred at this time. Formerly, as we have seen, its distinguishing mark had been connection with a name; it had been "Houghton's Regiment," "Warburton's Regiment," and so on; now it was to have a local habitation and a name— a name which it held down to 1881, and which it is hoped it may ever continue to hold as a deeply-prized tradition.

The state of the country and its foreign relations at this time were such as to cause alarm, but English patriotism was awake in the country, and the leaders of the county of Nottingham, taking advantage of the usual gathering in the city for the races in August, 1779, called a meeting of the nobility, clergy, and gentry of the county, under the chairmanship of Sir Robert Sutton, to consider what could be done by the city and county to advance the service of the State.

A committee was elected, which, after discussion, unanimously resolved "That the chairman of this committee do write to the Secretary at War, transmitting to him copies of the resolutions of the general meeting and of the present committee, and do request him to move His Majesty to appoint some particular regiment to be recruited in this county with the assistance of the subscriptions entered into, and that His Majesty be graciously pleased to order such regiment henceforward to be distinguished by the name of the county."

The request was at once complied with. A depôt of the 45th marched to Nottingham "on recruiting service," and the people were assured that as soon as 300 men should be raised for the regiment in the county it should be distinguished by the title of "Nottinghamshire Regiment." As in addition to the usual bounty a sum of six guineas was paid from the county subscriptions to each recruit, many more than the stipulated number were speedily obtained, and the 45th received its county appellation, and regiment and county have been ever since in close connection. During the French revolutionary war the regiment received many hundreds of men from the county, nearly all of whom were volunteers from the Nottinghamshire Militia. Captain Lawson Lowe, in his history of the "Royal Sherwood Foresters," says :— "Few regiments under the Crown have, during the last 80 years, seen more arduous service than the 45th, and none have earned for themselves a higher renown. Should any be disposed to sneer at this bloodless record of a militia regiment, let them look at the honours upon the tattered colours of the gallant 45th, and remember it was by Nottinghamshire militiamen that those honours were won."

In 1781 it was ordered that every sergeant and corporal should carry a bullet mould and a ladle to melt lead in, with three spare powder horns for priming, and twelve bags for ball. It was also directed that an officer when dressed for duty should have his hair "queued," sash and gorget on, buff gloves, black linen gaiters with black buttons, black garters, and uniform buckles.

In 1782 the regiment was still stationed at Chatham and Rochester, with a company recruiting at Southwell. It was commanded by Colonel Haviland, and in the latter part of the year moved to Dover. Although this is the first year in which a chaplain appears in the regimental returns, one had been attached to the regiment from the first; this was the Rev. Robert Brenton, who served till the end of 1783, a period of 43 years.

In 1783 the 45th first appears in the army list with its
territorial title of " First Nottinghamshire Regiment," the
agents being Messrs. Cox, Mair & Cox.    In October the
regiment was at Newcastle-on-Tyne, and in the following
month at Tynemouth.    The strength at this time is given as
165 rank and file, though it is not easy to understand why it
should have been so weak, when only four years previously
it had received such a large accession of strength through its
connection with the county of Nottingham.

In the following year the regiment passed through Carlisle
to Dumfries, with detachments at Wigtown, Stranraer, and
Kirkcudbright, and on the 27th of July it embarked at Port
Patrick in the transports " Sisters Rodney " and " Duke of
Leinster " for Donaghadee, moving on to Armagh on the 14th
of August.    The disembarkation strength is given as :—
1 lieut.-colonel, 3 captains, 5 lieutenants, 4 ensigns, a quartermaster, a surgeon, and a surgeon's mate ; 16 sergeants, 20
corporals, 10 drummers, and 216 privates.    It was commanded by Lieut.-Colonel Dundas, and was inspected at
Armagh on the 4th of October by Sir Henry Calder, when it
showed a strength of 255 rank and file.

On the 1st of March, 1785, the headquarters were at
Waterford, with detachments at Fort Dungarvan and Charles
Fort.    During this year the light company was equipped
with buff belts instead of black, and long black gaiters were
adopted in place of the half-gaiters then worn.

On the 2nd March, 1786, the regiment left Monkstown for
the West Indies, having then eight companies and a strength
of 20 officers, 14 sergeants, 9 drummers, and 303 rank and file.
17 officers, 12 sergeants, 48 women, and 36 children were left
behind in England.    The regiment landed at Grenada, and
during its stay there was actively employed.    Lieutenant
Guard was wounded in an attack on one of the islands.

As illustrative of how things were managed in those days,

it is worthy of note that three sons of the Adjutant, Lieutenant Darling, were enlisted as recruits at the headquarters. These were probably children, thus following a practice common in the navy.

On the 9th of October the Quartermaster Lieutenant Richardson died at Grenada. The year 1787 was a most disastrous one for the regiment, the deaths averaging 9 to 14 monthly out of a strength of about 340.

In 1788, on the 8th March, Lieutenant Huthwaite died, and during the year the Adjutant and Quartermaster exchanged appointments. In the following year two officers of the regiment died—Lieutenant James Robertson on the 7th of April, and Captain William Sutherland on the 2nd of May.

During 1790 the strength of the regiment was reduced to 268 rank and file. On the 9th of June Lieutenant W. H. Clinton, from the 7th Light Dragoons, was appointed to a company in the 45th, from which he exchanged on the 14th of July to a lieutenancy in the 1st Foot Guards. He died in 1846 as General Sir W. H. Clinton, G.C.B., Colonel of the 55th Foot, and Lieut.-Governor of Chelsea Hospital.

In June, 1791, Captain Thornhill Heathcote died at Grenada, and in December the corps was reinforced by 100 men from home.

Among the orders for this year was one directing that all recruits on arrival at headquarters were to be supplied with a uniform dress, to consist of a scarlet jacket, trousers, and round hat, etc. Field officers were to wear two epaulettes to distinguish their rank, and officers of flank companies to wear wings. The sergeants were given a short pike instead of a halberd, and soldiers serving in the East or West Indies were to wear round hats, broad leaved, and of the Cromwellian shape.

In January, 1792, the regiment was reinforced from

England by 10 sergeants, 10 corporals, and 170 rank and file, bringing its strength up to 337 rank and file.

Sickness still continued, and on the 2nd of April, 1793, Lieutenant Norman died, and on the 1st May Lieutenant and Adjutant Nicholls.

It is said that during this year a part of the regiment returned to England for service under Sir Charles Grey. This, however, seems hardly likely, and it would be more probable that a part of the regiment either volunteered, or was detached, to join Sir Charles Grey on his arrival in the West Indies, as no mention is made of the 45th in the lists of the regiments which sailed with the expedition from Portsmouth.

War had been declared against France on the 1st of February, and the safety of our own West India Islands, to say nothing of possible attacks on those of the enemy, would naturally have militated against the return of any regiment.

In March, 1794, Ensign Christopher Reve died, and in April the remains of the regiment, amounting to only 5 officers, 15 sergeants, 20 corporals, 8 drummers, 28 privates, 14 women, and 14 children embarked on the ship " Ulysses " for England, and, under command of Lieut.-Colonel Nicholls, landed at Portsmouth on the 28th of July, occupying Hilsea Barracks.

In August, 1794, the regiment proceeded to Guernsey, and in September its strength was returned as 10 companies of 237 officers and men. In November the headquarters were returned as at Hilsea, and in October orders had been received at Bristol that a party of the 45th was expected to arrive there by sea, possibly from the West Indies, *en route* for Hilsea. In December the strength of the regiment at Hilsea was returned as 686 officers and men. This rapid rise in strength is said to have been due to drafts from the prison

ships, but it is more likely that returns from the West Indies
and ordinary recruiting were the causes of the augmentation.

The increase in strength was also doubtless due to the
approach of foreign service.    The losses the troops had
suffered in the West Indies from the enemy as well as from
sickness demanded reinforcements, so on the 1st February,
1795, we find the regiment at Yarmouth, in the Isle of Wight,
preparing to embark for the West Indies, 678 strong.    The
embarkation took place on the 14th of February.

There was at this time little difference between the
accoutrements worn by the regiment and those worn for the
previous 35 years.    The "Kerein Kuller," with its stiff,
upright feather and white braiding, was still worn, as well
as the square skirted red coat, buttoned back at the tail to
show the white breeches and black leggings.    The hair was
still powdered and queued; the officers wore gorgets, ruffled
shirts, and little epaulettes of gold or silver, according to their
facings, those for the 45th being silver.    The sergeants still
carried the half-pikes; the drums were of wood, and the
colours destitute of embroidery, and about twice the size of
those now carried.

B

# CHAPTER III

1795 to 1807—West Indies; continued heavy mortality—Affair of the "Windsor"—Ireland—Disastrous expedition to Buenos Ayres.

ON the 5th of April, 1795, the regiment landed at St. Pierre, in the island of Martinique, having left, *en route*, in hospital at Barbadoes, 1 officer, 4 sergeants, 3 corporals, and 69 privates. Sickness continued rife; on the 29th of May Lieutenant James Pretty died, on the 28th of July Quarter-master Christopher Darling, on the 5th of August Major Haviland, and on the 4th of November Captain Sandreth.

In 1796 the regiment moved to Dominica, with detachments at Martinique and Iles des Saintes. A statement is made that during this year "the remains of the flank battalion that accompanied Sir C. Grey to the West Indies in 1794 were drafted into the 45th." This may probably refer to the return to the regiment of a detachment which remained behind when the regiment returned to England in 1794, as no trace can be found of any such regiment leaving England.

For the next three years the regiment led an uneventful life, but suffered terribly from disease. In 1797 and 1798 no less than 13 officers died, namely, Lieut.-Colonel Frazer; Captains Morrison and Hutchinson; Lieutenants Richardson, Eyre, Reynolds, Andoe, Cavanagh, and Mackay; Ensigns Cahill and Bustard; Surgeon Belfrage and Assistant-Surgeon Wilkins; and by 1799 the strength had fallen to 356 rank and file.

In May, 1800, the regiment moved to Iles des Saintes, and

there received 42 volunteers from the 38th and 43rd regiments, as well as upwards of 50 recruits from home, making up the strength to 34 sergeants, 34 corporals, 12 drummers, and 327 rank and file.

In March, 1801, the regiment moved to St. Kitts, where it shortly received orders to return home. The embarkation took place on the 4th May, on the transports "Alfred," "Aurora," "Lady Shore," "Mary," and "Windsor," the strength being only 12 officers, 31 sergeants, 23 corporals, 12 drummers, and 88 privates, no less than 202 men having volunteered for different regiments in the West Indies.

On board the "Windsor" were 150 French prisoners, and as a guard to them Captain Gwyn, Lieutenant Bond, Quartermaster Thresher, 14 non-commissioned officers, and 16 privates were detailed. All went well until the third or fourth night of the voyage, when the "Windsor" parted company with the rest of the squadron, and found herself alone.

On the next night, at 12 o'clock, the quartermaster, who was in charge of the military watch, went below himself to call his relief, Captain Gwyn; probably some of the sentries followed his example, for the Frenchmen, seeing their advantage, fell upon and overpowered the rest of their guard. They then seized the arm chest on the quarter-deck, secured the officers in their cabins, and took complete possession of the ship. That such an event had been considered possible is evident from the fact that representations had been made, previous to sailing, as to the weakness of the guard.

The Frenchmen then shaped their course for Boston, where the ship arrived on the 5th of July. The officers and men of the 45th returned to England via Halifax, Nova Scotia, and arrived at Plymouth on the store ship "Camel" in December. No one appears to have received any censure in connection with the occurrence, as the report made of the circumstance by Captain Gwyn was considered by H.R.H. the Commander-in-Chief as "perfectly satisfactory."

The regiment landed at Portsmouth on the 5th of July, 1801, and marched the next day for Horsham. In September it moved to Winchester, arriving there on the 12th, but left again on the 5th of November for Fort Monckton and Portsmouth.

On the 5th of April, 1802, the regiment embarked for Ireland on board H.M.S. "Revolutionaire," but was at once transferred to H.M.S. "Assistance," in which ship it reached Kinsale on the 14th or 15th, about 100 strong. After a short stay at Fermoy it returned to Kinsale, where it remained till the 5th of August, 1803, on which date it proceeded to Limerick.

Up till this time it had been the custom for the colonel and the majors to possess companies as if they had been captains. This arrangement had no doubt to do with their emolument, as pay, being small, was made up by means of perquisites and allowances. This plan was now abolished; Captain-Lieutenant Cunningham obtained the colonel's company; Captain Nicholson, brought in from half-pay, took the lieut.-colonel's company; and Captain and Adjutant Smith succeeded to the major's company.

Early in 1804 Colonel Montgomery quitted the regiment to take up an appointment on the staff in the West Indies, and Lieut.-Colonel Guard succeeded to the command. On the 20th of June the corps, leaving its light company attached to a light brigade formed at Limerick, marched to Ballyshannon. On the 1st of October, orders having been given to raise a second battalion at Mansfield, a party of officers, non-commissioned officers, and men left Ballyshannon for the purpose. The establishment of the new battalion was fixed at 43 non-commissioned officers, 20 drummers, and 800 rank and file.

While quartered at Ballyshannon the regiment lost three officers through the deaths of Ensigns Highmore, Osborne, and Williamson.

The rank of captain-lieutenant was abolished this year.

In July, 1805, the headquarters were at Enniskillen, moving in August to Kildare, where it formed part of an army of 15,000 men assembled for instruction under Lord Cathcart, commanding the forces in Ireland.

On the 27th of December the regiment marched to Monkstown, near Cork, and embarked on board the "American" for the Downs, where a force under Lord Cathcart, destined for North Germany, was to rendezvous. Napoleon's victory at Austerlitz, however, put an end to all idea of Continental operations, and the regiment was landed on the 14th of January, 1806, at Deal and Ramsgate, and quartered in Brabourne Lees Barracks.

The second battalion was at this time at Chelmsford, about 200 strong, and from it, in July, 20 volunteers joined the 1st battalion, as well as 19 volunteers from the 85th regiment.

Sir John Moore, who commanded the district in which the regiment was at this time quartered, appears to have given it special commendation after one of his inspections, and announced his intention, on parade, of recommending it for " immediate, or any, service."

In June the regiment was at Shorncliffe, under Major-General Hill, but on the 14th and 15th of July it marched for Portsmouth, where it arrived on the 24th and 25th, and at once embarked for service, the nature of which was kept a profound secret. However, it was for some reason shortly after disembarked, and encamped at Buckland under Brigadier-General Sir Samuel Auchmuty, an old 45th man. The regiment was under the command of Lieut.-Colonel Guard, and its strength at this time was 649 rank and file.

The disastrous result of the attack upon Buenos Ayres by

Sir David Baird and Sir Home Topham caused an expeditionary force for the River Plate to be fitted out during this year under Sir Samuel Auchmuty, as well as a force for the invasion of Chili under Brigadier-General Robert Crauford, another old 45th man; and in August the regiment embarked to form part of the latter expedition.

By the end of September the expeditionary flotilla was assembled at Falmouth. The troops embarked consisted of the 1st battalions of the 5th, 36th, and 45th regiments, the 88th, and five companies of the 95th regiment; two squadrons of the 6th Dragoon Guards, and two companies of the Royal Artillery. The convoy was commanded by Rear-Admiral Murray, who hoisted his flag on the "Polyphemus"; and the transports which carried the 45th were the "Brunswick," the "Fame," the "Francis and Eliza," the "Indefatigable," the "Lady Delavel," and the "Ariel."

The embarkation strength of the regiment was 892 non-commissioned officers and men.

After the expedition had sailed, the news of Beresford's surrender reached England, and orders were sent after Admiral Murray altering the destination of the expedition, and ordering it to proceed to the River Plate. These orders, however, did not reach till the expedition had arrived at the Cape.

The equipment at this time had undergone but little change; the sergeants, in addition to the sword, still carried the half-pike. The soldiers on the line of march carried from 75 to 80 lbs.; this included the knapsack and complete kit, greatcoat, blanket, flint-lock musket, and accoutrements, with sixty rounds of ball cartridge, clothing, canteen, three days' provisions in the haversack, camp kettle and bill-hook. The expedition, which sailed on the 30th November, was detained at Porto Praya, Santiago, one of the Cape de Verd islands, for about a month, in consequence of rumours that a French squadron of superior strength was in the vicinity; and the troops were employed in erecting works for the security of the harbour in case of attack.

About the end of January, 1807, the expedition reached the Cape, where it landed, and on the 22nd of March the regiment was inspected by General Crauford, who expressed his entire approbation of its appearance and condition. The aptitude of the 45th for settling down on board ship had, by this time, become an established feature in its character. The General bestowed marked praise on the state of the ships conveying the regiment, and in his numerous inspections had highly commended their cleanliness and good order. One of the transports, the "Fame," on board of which the troops were commanded by Major Gwyn, was held up as a model, and officers of other corps were directed to visit her in order to inform themselves of the 45th system, which is stated in the records to have "approached as near as well could be to the naval system."

The regiment sailed again from the Cape on the 6th April, still in ignorance of their actual destination, though probably there was a strong suspicion that they were bound for Buenos Ayres, as on the 7th April Captain Whittingham, one of the D.A.Q.M.G.'s of the expedition, writes in his diary:— "Yesterday the Admiral made the compass signal to steer north-west during the night. This has decided my opinion as to our present destination. We are certainly going to St. Helena, and thence to Buenos Ayres." The expedition reached St. Helena on the 22nd of April, and sailed thence on the 26th, anchoring on the 27th of May, ten or twelve miles east of Maldonado.

On the 1st of June the "Flying Fish" came from Monte Video, bringing news from Rear-Admiral Sterling on the "Diadem" that he and Brigadier-General Auchmuty had arrived at Maldonado on the 5th of January with about 3000

men, finding there the remnants of Beresford's forces, with
which his own made up about 5000 men, including the 17th,
20th, and 21st Dragoons, some artillery, the 38th, 40th, 47th,
54th, and 87th regiments, and some companies of the 95th
Rifles.

The possibility of capturing Monte Video with this force
was soon decided on.

Monte Video, though protected on the land side by
ramparts and on the sea side by works, was not a strong
place when attacked by land and sea. The town is built at
the end of a peninsula, and its land face is scarcely a mile
in extent, making investment on that side easy. On the sea
side its strength lay chiefly in the shallowness of the water,
making the attack by heavy ships almost impossible, but at
the same time the investment by light ships and boats was
not difficult. The naval force on the occasion consisted of
the 64-gun ships "Diadem," "Raisonable," "Ardent," and
"Lancaster"; the frigates "Leda," "Unicorn," and
"Medusa," with some sloops and gun-brigs.

The whole force assembled near the island of Flores, and
on the 16th of January a landing was effected near Carreta
Point, about seven miles east of Monte Video. On the 19th,
the army moved forward in two columns, while the smaller
craft and boats of the squadron crept along the beach on the
left flank in support.

Auchmuty soon invested the town, opening with his
batteries on the works on the 25th, and on the 5th of February
it was carried by assault, with a loss to the conquerors of 140
killed and 350 wounded.

Monte Video being thus secured, preparations were made
for the next step, the capture of Buenos Ayres. A considerable number of vessels had fallen into our hands at Monte
Video, including some gunboats, which, being of light
draught, were equipped for operations up the river.

The approaches to Buenos Ayres by water were closely and carefully examined, and in consultation with Colonel Pack and Major Tolley of the 71st regiment, both of whom had been in occupation the year before, plans were arranged for the attack.

In May Lieut.-General Whitelock, who had been sent out to take supreme command in the River Plate, arrived with a further reinforcement of 1600 men.

All this was news which Crauford's division were very ready to receive. The troops were in excellent condition notwithstanding the fact that they had been at sea, almost without a break, for nearly nine months. The system of the 45th had told, and the regiment had hardly a man sick.

On the 19th of June things began to move, and a division of Crauford's transports moved up and seized Colonia, the port opposite Buenos Ayres, as an advanced base.

On the 21st of June the Admiral shifted his flag to the "Nereid," and with the "Medusa" and "Thisbe," having the Naval Brigade on board, sailed up the river with the last division of the transports, leaving the five or six line-of-battle ships behind at Monte Video.

The completed part of the city of Buenos Ayres, at this time, filled the area of an equilateral triangle, having sides of about a mile and a half in length, the base resting on the river bank. It was built on the plan, common in America, of straight streets intersecting each other at right angles, dividing the buildings into blocks of about one hundred and forty yards square. The shore upon which the town abuts runs nearly north and south, while the two longest streets, running east and west, formed the perpendicular let fall from the apex of the triangle bisecting the base. At the point of bisection, enclosed between the two streets, was the great square and market place; and between it and the river

lay the bastioned citadel which had been held by General
Beresford when he surrendered in August, 1806.

The main objective in any attack would clearly be this
citadel; and, leading up to its capture, the possession of the
buildings forming the sea face of the town would seem a
natural step.  The strong positions defending the flanks of
this line were the Plaza del Toros, a mile from the citadel
to the north, and the Convent of the Residentia, about the
same distance to the south.

Gradually to the north, south, and west the houses ceased,
but the ground was still laid out in plots, stiff hedges taking
the place of buildings.

The houses were flat-roofed, with a parapet round each,
giving the defenders a terrible advantage by enabling them
to fire down into the streets from a position of comparative
safety.

A suitable landing-place was discovered at Ensenada de
Barragon, about forty-three miles below Buenos Ayres, and
here the " Nereid," with the naval and military commanders-
in-chief on board, cast anchor on the 24th of June.

It had been intended, as we have seen before, to use
Colonia as an advanced base, but these plans were changed,
and on the 27th of June Major-General Leveson-Gower was
despatched thither to carry out the evacuation of the place,
and bring the troops to Ensenada.   There seems to be some
question as to whether Leveson-Gower carried out his
orders as efficiently as he might have, especially in the
matter of transport of the horses, as it is clear that later,
through this or other causes, horses were very scarce.

The transports were arranged in three divisions.  Captain
Thompson of the "Fly," who knew the landing-place, was
directed to lead the 1st Division, having with him the
"Dolores" schooner and four gunboats; Captain Palmer in
the "Pheasant" was to lead the 2nd Division, with the

"Haughty" and two gunboats; Captain Prevost in the "Saracen" was to bring up the 3rd Division, and Captain Corbet was to superintend the landing.

The troops were arranged in five brigades, as follows:—

Three batteries of Light Artillery under Captain Frazer.

The 5th, 38th, and 87th Regiments under Sir S. Auchmuty.

The 17th Light Dragoons, the 36th and 88th Regiments, under Brigadier-General the Hon. Wm. Lumley.

Eight companies of the 95th Rifles, and nine Light Companies, one furnished by the 45th, under Brigadier-General Crauford.

Four troops of the 6th Dragoon Guards, the 9th Light Dragoons, the 40th and 45th Regiments, under Colonel the Hon. T. Mahon. All the Dragoons were dismounted, with the exception of the 17th Light Dragoons under Lieut.-Colonel Lloyd.

In addition, 200 of the Naval Brigade, under Captains Rowley and Joyce, were landed.

The strength of the whole force was 7822 rank and file, including 150 mounted men, with 18 guns and 206 horses and mules for their conveyance and that of the small-arm ammunition.

Soon after 9 a.m. on the 28th of June, the first boats, with Crauford's brigade, landed about a mile to the westward of the small fort, which the enemy had evacuated, taking the guns with them. The remaining brigades followed, and the whole force was landed without opposition.

At the landing-place the country was low and swampy, and a bog, knee-deep in mud, extended for four or five miles inland, beyond which the ground was higher. There were two routes leading to Buenos Ayres, one along the shore, almost an impenetrable morass at this time of year; the other from five to seven miles inland, on higher ground, which, though crossed by several streams, was quite passable, and was the ordinary route used by the inhabitants between the landing-place and Buenos Ayres.

The information possessed by the Generals seems to have been fairly complete. They knew that, after marching about 25 miles, they would reach the village of Reduction, only about 3 miles from the shore, where communication with the fleet would be easy, and where supplies could be landed. Colonel Pack and Major Tolley of the 71st had landed near Reduction the year before, and occupied the village. Knowing this, General Whitelock at once sent forward the Light Brigade, with the 38th and 87th regiments, and two three-pounders, under Leveson-Gower, to occupy some heights about five miles to his front, for the purpose of covering the landing, which continued all day. By the evening, however, only four six-pounders and four three-pounders had been landed, the greater part of the artillery, horses, and stores being still on board.

The next morning the General joined Leveson-Gower, having landed nearly all his troops, but with only a few horses and no transport. Leveson-Gower was then sent on with the advanced guard, consisting of the 36th and 88th regiments; while Colonel Mahon's Brigade, with the exception of the 45th, was left behind to cover the rear and bring up the artillery and stores.

We are struck by a certain amount of haste, as well as unnecessary vacillation, in these early changes in the original organisation and plans; but in his despatch the General put forward as an excuse the necessity of procuring cover and fuel. Nevertheless, the despatch reads like an early note of failure, and reading between the lines of the evidence given at the subsequent court-martial one gains the impression that there was, at least, a want of the necessary cordiality in the relations between the General-in-Chief and the other superior officers.

The General appears to have formed the general plan of getting his army round to the west of the city and taking up a position on its north-west face, with his left flank in communication with the navy, necessarily possessing himself for this purpose of the strong post of the Plaza del Toros; but he does not seem to have made this clear to the Admiral from the outset, and this officer consequently could, from the first, do no more than what his knowledge of the situation prompted.

He sent forward Captain Thompson in the " Fly," with the " Staunch," " Paz," and " Dolores," to open communication with the army when it should have reached Buenos Ayres, while he himself—evidently doubtful of the wisdom of their losing touch with the fleet—moved up along shore endeavouring to recover communication.

On the 1st of July he was near Reduction, about 8 miles below the city; here he sent the provision ships close in shore, and despatched Lieutenant Bligh to find the army. This officer found the General at Reduction on the 2nd of July, from whom he brought back letters asking for supplies of bread and spirits, and informing the Admiral of his intention of marching to turn the river Chuelo, and gain the westward of Buenos Ayres. He further requested him to send some of his squadron to meet him there with the store ships and reserve artillery.

The army at this time appears to have been suffering considerably from fatigue and want of provisions, although they had been four days in covering some thirty miles or so.

At Reduction the General learned that the enemy had fortified the opposite bank of the Chuelo, and intended to make a stand there. Accordingly, he determined to turn his position, and for this purpose struck to the left, marching in two columns; he himself commanded the left column, and General Leveson-Gower the right. Gower was ordered to find a ford which was reported to be practicable, and, if possible, to cross the river and make his way to the north-west of the city and reopen communications with the navy. Whitelock himself intended to cross the river higher up, and rejoin Leveson-Gower near the city. At the same time orders were sent to Colonel Mahon, who was now bringing up the greater portion of the artillery, the 17th Light Dragoons, and the 40th regiment, to halt at Reduction till further orders. Leveson-Gower, on his way to the ford, met and dispersed some of the enemy's cavalry; he then forced the passage of the river, and sent Crauford forward to seize a height at the apex of the triangle formed by the city, while he got his guns across the ford, which was deep and muddy. Crauford sent back word that a considerable force of the enemy's infantry and artillery were making for the same hill, and requested permission to attack them.

Finding Lumley's Brigade fit to advance, Leveson-Gower himself went on and joined Crauford's Brigade. The enemy was found in a strong position among the enclosures, and considering that he would lose fewer men by attacking at once, he ordered Crauford to charge the position with the bayonet, which was carried out with such vigour that the enemy was completely routed in a few minutes, leaving ten guns in the hands of the assailants. The light company of the 45th took a prominent part in this gallant attack.

By the time the troops had re-formed and a guard been set over the captured guns, it was nearly dark; Lumley's Brigade had arrived, and the division occupied a secure position for the night. The place was known as the Coral de Miserere, or sometimes as the Miserere, and though not sufficiently elevated to overlook the whole town, much of which was hidden by the thick hedges, gave a good view of all the higher and more prominent buildings.

The long detour to the left taken by Whitelock's column prevented its joining the right column until the afternoon of the 3rd. By this time the relations between Whitelock and Leveson-Gower appear to have become thoroughly strained. With whom the blame lay it is difficult to say, but the fact is well established, as, on the 4th, Whitelock told Leveson-Gower that he was "his declared enemy," and threatened to suspend him from duty.

As the troops came up they took post right and left of the positions already occupied by Crauford's and Lumley's Brigades: the 45th was on the extreme right, the 6th Dragoon Guards next, then came the Light Brigade, the 88th, 36th, 5th, and 87th regiments, with the 38th on the extreme left.

As we have already seen, Whitelock's original intention had been to take his whole army round to the north-west side of the city and open communication with the fleet; consequently, the position taken up by Gower was not the one he had intended. Probably, also, every one in the army was aware that, whatever else was done, either the post to the north alone, or that to the south as well, were to be attacked, and that the whole army would probably move round clear of the town on to the Plaza del Toros. Gower, however, seems to have had views of his own, and to have hatched a detailed plan of attack which, in spite of the coolness existing between them, he was ready enough to press upon the Commander-in-Chief as a substitute.

General Leveson-Gower, soon after daylight on the 3rd, sent Major Roche, Brigade-Major to Lumley's brigade, with a flag of truce to General Ellio, the Spanish commander, summoning him to surrender. The answer was returned desiring to know the terms asked.

The Spanish General was probably aware that the defeat of his troops on the previous night had been effected by a part of one wing of the army which was now in position overlooking the city, and that another column with artillery was at Reduction ready to advance; he must also have seen the war ships passing in front of the city.

There can scarcely be a doubt that, if mild and easy terms had been offered, there would have been a surrender and an instant cessation of hostilities; but this was not to be, for in accordance with previous instructions from General Whitelock, Leveson-Gower put forward terms of the utmost severity, demanding, amongst other things, that " all persons holding civil offices dependent on the Government of Buenos Ayres become prisoners of war," and that all public property of every description be delivered up to the British commander. The Spaniards had but one answer to make, and they made it—defiance.

No wonder that the court-martial which sat subsequently found "that the said Lieut.-General Whitelock in making such an offensive and unusual demand—tending to exasperate the inhabitants of Buenos Ayres, to produce and encourage a spirit of resistance to His Majesty's arms, to exclude the hope of amicable accommodation, and to increase the difficulty of the service with which he was intrusted—acted in a manner unbecoming his duty as an officer, prejudicial to military discipline, and contrary to the Articles of War."

So on the afternoon of the 3rd of July it was certain that an attack was to be made, and the unfortunate general in command gave up his own perfectly rational plan of attack for the most ill-advised one devised by his subordinate. This was, briefly, to break the whole army up into wings of regiments, and to march in about fourteen columns down as many streets, straight upon the river. The objective was not to be the citadel; consequently, the two central streets were to be avoided. If penetration towards the river could not be effected because of the resistance encountered, the columns were to work away towards the right and left and to look upon the Plaza del Toros and the Residentia as their bases; the 38th and 87th being told off to capture the former, and the 45th the latter.

The columns were, therefore, to pass down the streets of a town where they stood a good chance of being decimated by fire from the roofs of the houses, only for the ultimate purpose of falling back upon positions which they could most easily have gained without entering the town at all.

Moreover, the dangers from their own fire in the intersecting streets promised to be so great that orders were given for the advance with unloaded muskets, and "no firing was to be allowed on any account"; while two companies of the 88th, who were slow in unloading, were ordered to knock the flints out of their muskets, and so entered the town practically unarmed. This order was the only one on which the General was acquitted at the subsequent court-martial. "The court was anxious that it might be understood that they attach no censure whatever to the precautions taken to prevent unnecessary firing during the advance of the troops to the proposed points of attack, and did, therefore, acquit Lieut.-General Whitelock of the said charge." With such a hopeless scheme, what chance could there be of success? Nevertheless, the die was cast, and the fourteen columns were ordered to advance down the streets opposite to them. In circulating the plan the General himself condemned it in sufficiently strong terms. He said—" The refusal of the Spanish General to listen to terms, and the state of the army from fatigue and bad weather leave but little choice as to the mode of accomplishing our purpose; otherwise, I should assuredly be disposed to adopt one equally calculated to secure to us

c

possession of the place without the probable chance of so much blood being spilt."

The astonishing reflection is that he must have known that he was referring to British blood only, for with no British firing, the Spaniard was safe on his housetop.

The signal—a gun from the centre—was fired at the appointed hour, and all the columns moved forward. The 38th marched in one column on the extreme left, with the 87th in two columns on their right; then came the 5th Fusiliers, the 36th, and 88th, each in two columns, marching down contiguous streets, the right wing of the 88th taking the street next to the one which led direct to the north, or left side of the Grand Square and citadel. The two centre streets were unused, the light battalion marching down the next two streets on the right in two columns; then came the 6th and 9th Dragoons together in the next street; then the left wing of the 45th under Major Nicholls, with the right wing under Lieut.-Colonel Guard on the extreme right.

It will be remembered that Colonel Mahon, with the bulk of the artillery, had been left at Reduction; orders were sent, which reached him on the morning of the 5th, to halt "at a safe distance" from the bridge over the Chuelo. His total force consisted of about 1800 men, including the 17th Light Dragoons, the 48th Regiment (excepting its light company), one dismounted troop of the 9th Light Dragoons, a detachment of the 45th under Major Gwyn, one company of the 36th, a detachment of the 88th, about 50 men of the artillery, and 200 seamen; 4 six-pounders and 2 howitzers. One of the most extraordinary circumstances connected with this ill-advised attack is the inactivity of this column when there was plenty of time to have brought it up to join in the desperate service before the fourteen small attacking columns.

Let us now turn to the progress of the attack. The 38th

regiment, under Major Nugent, from its position on the extreme left, did not meet with much resistance until it came under fire from the Retiro and the Plaza del Toros. The former, however, was captured without much difficulty, and Major Nugent turned the fire of a 12-pounder, which he found there, upon the Plaza del Toros, which surrendered after a little while with 400 men; the place was soon occupied, and was held, together with the Retiro, by the 38th and a reinforcement of the 87th with Sir S. Auchmuty. The 87th advanced a considerable distance without meeting any opposition; it was still dark, however, when they were assailed by a discharge of grape on their front. The column pushed on, when severe musketry fire was opened on them from the Plaza del Toros on their left. One wing fell back upon the streets to the right, and, reaching the river, occupied a large house, and was subsequently joined by the left wing, which had fallen back in the same way.

Sir S. Auchmuty, who commanded this column, presently saw the colours of the 5th Fusiliers planted on the Convent of Santa Catalina, to his right, and sent parties to secure his rear by possessing themselves of the houses there, from the roofs and windows of which the enemy were firing on him. While this was going on, new firing was heard on the left, and as soon as the houses in the rear were evacuated, the 87th opened communications with the 38th, whose fire they had heard, and who were just completing the capture of the Plaza del Toros; and, joining them in its occupation, they somewhat withdrew their support from the 5th Fusiliers. This regiment must naturally have expected earlier and stronger resistance than that encountered by the 87th, as the streets it had to traverse were longer, and it was so much nearer to what was considered the centre of the defence. The left wing, advancing rapidly at the charge, reached the river without opposition, the enemy abandoning four guns

in the street, which, however, they had previously spiked;
the church of Santa Catalina and several houses near were
seized, and the colours of the regiment hoisted.     Between
nine and ten o'clock Major King, who commanded the wing,
noticed that the firing at the Plaza del Toros had ceased, and
that the British colours were displayed there.

The right wing of the 5th, under Lieut.-Colonel Davies,
also reached the river unopposed, and, occupying some of
the houses there, hoisted the King's colours upon them; he
then assisted in occupying the church of Santa Catalina,
and subsequently went himself to the left to obtain further
orders from Sir S. Auchmuty.   Sir Samuel came back himself, examined the post, and bid him hold it.   About one
o'clock a message came from Brigadier-General Lumley,
with the 36th on his right, asking him to advance instantly
to his support.   When Colonel Davies, with three companies
of the 5th, joined General Lumley he found they were all
ordered to fall back to the left on the Retiro and Plaza del
Toros, which was accordingly done.

The 36th regiment, under Colonel Burne, had been moved
close up to the town before the signal was made, and its two
wings moved down the two streets to the right of the right
wing of the 5th.     They found the streets they had to
traverse much broken up, and fire was soon opened upon
them, but in spite of it all they pushed on at a good pace, and
succeeded in penetrating to the last cross street before reaching the river bank.   Some of the houses facing the river
were with difficulty broken into and occupied.   The cross
street in rear of the position was now enfiladed by two guns
from the great square, and two guns brought out of the
citadel opened fire on the front.   Colonel Burne took a
detachment and made a dash for the two guns astride the
fort, which, after some loss, he captured, spiking one of them
with his own hands.

The position of this column, fired on from front and flank, now became somewhat precarious, as no support was forthcoming. Soon the Spanish General Ellio came with a flag of truce, demanding the surrender of the 36th, saying that the 88th had already surrendered. This was peremptorily refused, and General Lumley sent to Colonel Davies of the 5th, who had meanwhile driven away the guns and infantry which had been annoying him from the great square, to come to the support of his left wing. General Ellio again summoned Lumley to surrender, and Lumley again refused.

In General Lumley's account of his own movements he says :—" I had now received intelligence by a pencilled note from Captain Watson that Sir Samuel Auchmuty was in possession of the Plaza del Toros, and that he recommended my retiring along the beach and joining him on the heights of the Retiro. I wrote an answer on the back of the same note : ' I am still in possession of my post, cannot Sir Samuel support me ? ' This note never reached its destination. It was now near two o'clock; we had been engaged in this unequal contest for about six hours, our numbers much reduced, men and officers falling fast; forbid in the first instance in our instructions to advance against the fort and square; all the ammunition being expended, and well aware that I must ere long be completely surrounded and overpowered by numbers, I deemed it most advisable for the good of the service, instead of a fruitless resistance, to fall back and reinforce Sir S. Auchmuty with the remains of the 5th and 36th regiments then with me. I accordingly retired along the beach, still exposed to a heavy fire of grape and round shot from the fort, and with some additional loss, between two and three o'clock, I joined and put myself under the orders of Sir Samuel Auchmuty with the two regiments above mentioned."

Thus the 87th, 5th, and 36th regiments found themselves, after suffering heavy loss in their dangerous march through the streets, just where they would have been, practically untouched, had they been kept outside the town in rear of the 38th.

The left wing of the 88th, which was on the immediate right of the 5th Fusiliers, and was commanded by Major Vandeleur, moved in sections with a front of seven files, as wide a face as the street permitted. Before they had advanced far, fire was opened on them from the house-tops and windows on all sides. Pushing their way through, however, they carried a breastwork and ditch thrown across the street, and finally broke open and occupied a house facing the river. Here, however, little or no cover was to be found, and, unsupported and losing heavily, nothing remained for this wing of the 88th but surrender.

Lieut.-Colonel Duff, who commanded the right wing of the 88th, " had so bad an opinion of the attack in his own mind that he left the colours at headquarters, fearing they might be taken." He was ordered to take possession of a church, the dome of which could be seen over the town, which was some streets back from the objective—the sea. This wing met with no opposition till it reached the gateway of the church; then fire was opened on all sides, and 30 men fell at once. It was found impossible to break open the doors, and it was necessary to push on in search of some tenable position. Three houses in the direction of the citadel were seized, and held for four hours; when, being surrounded and reduced to some 80 or 100 men, they had no option but to lay down their arms.

The total result, therefore, of the attack by the left wing of the army was the loss of the 88th regiment, and a retreat of three decimated regiments from the ground they had gained to the point from which they ought, in the first instance, to have advanced.

The story of the right wing of the army is much that of the left, but possibly without the blame attaching to the commanders which cannot be withheld from those of the left wing.

Unquestionably the greatest success and distinction which fell to the lot of any corps present came to the gallant 45th and its leaders.

Brigadier-General Crauford's command, which was composed of the 95th and the light battalion, was divided into two columns, each accompanied by a 3-pounder gun; the left column, 600 strong, under Colonel Pack of the 71st, who knew the town, and the right column, about 540 strong, under Crauford himself. Crauford's orders were to penetrate, if possible, quite down to the river, and there occupy "any of the high buildings as near as possible to the market place."

Colonel Pack penetrated, unopposed, except for a few random shots at long range, right down to the river, where, hearing firing on his left, he determined to take ground in that direction and commence the attack, believing that he would receive support from Crauford's column and the 45th. Accordingly, he subdivided his force into two parts, and advanced by two parallel streets upon the square.

"I was always," says Colonel Pack, "under apprehensions that we were unprovided with means equal to overcome the defences of the place, and I was soon convinced that I had entered upon a contest the most unequal, perhaps, that was ever fought, for I had scarcely approached under the Franciscan church when I lost, by the fire of the enemy—almost invisible, and certainly by us unassailable—the officer and almost the whole of the men who composed my first division, who were volunteers from the different companies; the officer and nearly half the next company, and so in proportion of the others who composed my division.

"Finding it impossible to penetrate to the object of attack, which I conceived to be the square or fort, or gain any advantageous position in the neighbourhood, I thought it right to desist, and enquire the success of the division which had gone by the approach on my left. With this intent I withdrew the remains of mine to the cross street by which we had advanced, and which in a great degree protected them from the fire we had been under. I had scarcely done so when I learned the failure of my other column also; and, on going into the street, I found the men retiring, and met Colonel Cadogan in a little time. He was excessively distressed, and emphatically assured me that he and his men had done their duty, but really they had not the means of succeeding. I should mention that the gun which was attached to our column had accompanied him; every man and horse had been killed or wounded at it, and the gun was lost. I could readily believe Colonel Cadogan's statement, and directed his men to form on line with mine, and I went myself to reconnoitre the buildings forming the bottom of the square, in which the Jesuits' college was situated, but though well acquainted with them, I found it impossible to find an entrance there. On returning to Colonel Cadogan I intimated my intention of proceeding to the Residentia. Some of his men had forced open two houses, but I conceived them to be of no consequence, except that they gave shelter to the wounded men. Colonel Cadogan, however, deprecated the idea of giving up the ground we had gained with so much loss, and as Brigadier-General Crauford was momentarily expected up, I allowed him to remain in the position the troops were then in."

Colonel Pack then went himself to meet the Brigadier, and report to him the state of affairs.

Crauford led his column right through the town as far as the beach, without meeting anything but small, straggling parties of the enemy. Seeing the south-east bastion of the fort about 450 yards from him, he determined to advance along the beach upon it, and sent orders to Colonel Guard of the 45th to advance towards him by the street nearest to the beach. He had not proceeded far when he met Colonel Pack, who strongly advised him to fall back on the Residentia, whence Colonel Guard had first arrived with the Grenadiers of the 45th, bringing news of its capture by the regiment. Crauford, however, hesitated to retreat, and finally determined to occupy the church of St. Domingo, and hold his ground here. "I little thought," he says, "when I thus, to the best of my judgment, obeyed the orders I received, that I should have been abandoned without an effort of any sort being made to communicate with me. I had been directed to take certain posts and there wait for further orders, and this was one of the posts."

Until near noon Crauford had no reason to suppose that any disaster had befallen the army, though he necessarily knew that Colonel Cadogan was isolated, as he was; and he must have been uneasy at hearing nothing of the 6th and 9th Dragoons. Shortly afterwards, a Spanish officer approached bearing a flag of truce, and, informing him of the capitulation of the 88th, summoned him to surrender, which he peremptorily refused to do.

The entrance to the church was too narrow to admit the three-pounder, which had been left in the street; and about one o'clock a considerable force of the enemy approached for the purpose of capturing it. Crauford, feeling that his position was not tenable, conceived the idea that, by mixing with the enemy's force, the evacuation could be more securely effected, as the fire from the houses could not be poured indiscriminately on friend and foe.

But while he was calling his men together, Colonel Guard with his Grenadiers, and Major Trotter with some of the light troops, dashed out and threw themselves upon the enemy, who at once gave way; the fire from the houses, however, was so hot that Major Trotter was killed and about 40 of the men were killed or wounded in two or three minutes. This convinced General Crauford of the hopelessness of evacuation, and he held on.

After holding his post for about eight hours, without a sign of relief, the uselessness of further resistance became apparent, and he surrendered with about 600 men, including 100 wounded.

The dismounted cavalry column, which should have marched down the street on Crauford's right, passed into the town with one gun in its front and another in its rear. The road was difficult, and by the time the column reached the fourth square, the commander, Colonel Kington, and several officers, had fallen, and it was found impossible to advance further. After retiring, it was again ordered to advance, and succeeded in holding some houses, from which the British colours on the Plaza del Toros and the Residentia could be seen. The position, however, was isolated, and the column was unable to move to the support of any other part of the army. The 45th was on the extreme right of the army, and was divided into two columns; Major Jasper Nicholls took command of the left wing, and Colonel Guard of the right; while Major Tolley of the 71st was detailed to accompany Major Nicholls and show him the way. Neither wing met with any formidable opposition, but the left wing somehow crossed the right, and, taking a detour to the right, did not reach the Residentia till it was all but in possession of the right wing. The capture of the Residentia was easy, and Colonel Guard, having made all secure there, and hearing firing to his left, marched down the cross street towards it with his Grenadier company; having successfully charged down the street he came under a heavy fire, which compelled him to draw off his party towards the river, where he met Colonel Pack, with the result already related. Major Jasper Nicholls proceeded to possess himself of the houses adjoining his post with the seven companies left under his command. His anxiety concerning Colonel Guard was great when he saw the streets up which he had passed being occupied in great force by the enemy. On the other hand, his mind was more at ease so long as he saw the British colours flying on the church of St. Domingo, seven or eight hundred yards to the north. But when about three o'clock these colours were withdrawn his hopes of success began to fade. The enemy continually pressed on him, but his parties always drove them back with ease, and on one occasion captured two field-guns.

It was one o'clock on the afternoon of the 6th before any relief came, when Captain Whittingham, General White-lock's A.D.C., arrived with 40 Grenadiers; at the same time the enemy again approached to attack; the 45th went for them at once, and with very little loss thoroughly routed them, capturing two howitzers and their limbers.

"The men," writes Major Nicholls to General Whitelock soon afterwards, "behaved like Britons, and, your excellency may rely on it, will continue to do so. Our loss is very trifling, in all but seven or eight killed, including some of Colonel Guard's column, as many wounded, and five missing.

"Major Tolley led us on in very excellent style; my dispositions have been the result of our joint opinions, and we have no doubt of being able to retain this important situation during your excellency's pleasure. By foraging I have most easily subsisted my men, and I have enough for two days for those under my command. Our only want is an artillery officer and a few men to use the cannon with which we can completely command the beach and streets, or perhaps assist in any attack which your excellency may direct against the fort.

"I should not have written one line on these subjects had not Captain Whittingham particularly requested it. I trust your excellency will excuse the length of this detail.

"The enemy have lost in all, in this quarter, from fifty to a hundred men, and humanity has induced our men to make some prisoners.

"Major Tolley and myself have held very frequent conversations as to advancing, but, from uncertainty as to Brigadier-General Crauford's situation, the enemy's force—the streets being lined—the weakness and lowness of the intermediate buildings, with the great strength of the large churches, we have not yet thought it advisable to quit this post, which would not be tenable, in our opinion, if thus abandoned by the chief part of this division."

But while this letter was being written the admission of absolute failure was in course of consummation by the British. General Whitelock, at five o'clock on the evening of the 5th, received a letter at the Miserere from the Spanish General, Liniers, stating that he had captured eighty officers and upwards of a thousand men, and offering to surrender all the prisoners, as well as those of Brigadier-General Beresford's command taken the year before, provided the British re-embarked the whole of their forces, evacuated Monte Video, and quitted the River Plate entirely.

The General at this time knew little or nothing from his own officers of what had happened; but, early in the morning of the 6th, he went with his staff to the Plaza del Toros to gain information and confer with the officers there. Here he found the troops in full communication with the navy, which had landed a 24-pounder on the beach, and had brought up four gunboats, which were keeping up an effective fire on the town, being within point-blank range of the citadel.

From the Plaza del Toros General Whitelock replied to the Spanish General, declining his terms, but suggesting an armistice for twenty-four hours. Later on, however, after consultation with General Auchmuty and the Admiral, he came to the opposite conclusion, and determined to accept the terms offered. So, on the 7th, a definitive treaty was drawn up and signed, and hostilities ceased.

The gallantry and success of the 45th had been so marked that it was stipulated that it should quit the post it had so gallantly captured and defended, with all the honours of war. Accordingly, it marched through the streets with colours flying, drums beating, and bayonets fixed; taking the captured guns loaded, and with lighted matches.

General Whitelock in his despatch says:—"Now I should own the gallant conduct of the 45th, who, on the morning of the 6th, being pressed by the enemy near to the Residentia, charged them with great spirit, and took two howitzers and many prisoners."

By the 18th July the whole army was reassembled at Monte Video preparatory to embarkation, and, on the 30th December, the 45th disembarked at Cork 589 strong, having lost on the River Plate fourteen rank and file killed; two captains—Captains Greenwell and Payne; one lieutenant—Lieutenant Moore, who subsequently died of his wound; four sergeants, and forty-one rank and file wounded, and one missing. It had been seventy-three weeks on board ship, with the exception of the twenty days at Buenos Ayres, and the short landings at Porto Praya and the Cape. Major Nicholls was promoted to the rank of Lieut.-Colonel in the Royal York Chasseurs.

An incident which has always been held as a tradition in the regiment, but to which no exact date has ever been assigned, I consider must have happened probably between the date of the regiment's return from South America and its embarkation for the Peninsula, as it was during Colonel Guard's tenure of the command.    I give the story in the words of Colonel Campbell, who was adjutant of the regiment at the time: — "I remember well Colonel Guard, whose adjutant I was at the time, being most anxious that the 45th regiment—which he had for some years commanded—should be made light infantry, and also to have had them styled ' The Sherwood Foresters.'   He, however, for what reason I know not, failed in the objects he had in view.   Not long after, the 45th was brigaded in England for exercise with the 87th and 88th regiments.   Colonel Guard had constantly, and  much  to  his  annoyance—and  more  particularly on account of his recent failure—heard these corps called to attention by their appropriate local designations in place of their numbers ; but one day he could stand it no longer, and when Colonels Butler and Duff loudly and proudly exclaimed, ' Prince's Irish,' and ' Connaught Rangers,' he, in a very shrill voice, called out at the same instant, ' Nottingham Hosiers, attention ! '   His brother chiefs, who seemingly had not heard or understood  what he had said, looked all astonishment when the whole brigade burst into an irrepressible and unmilitary fit of laughter."

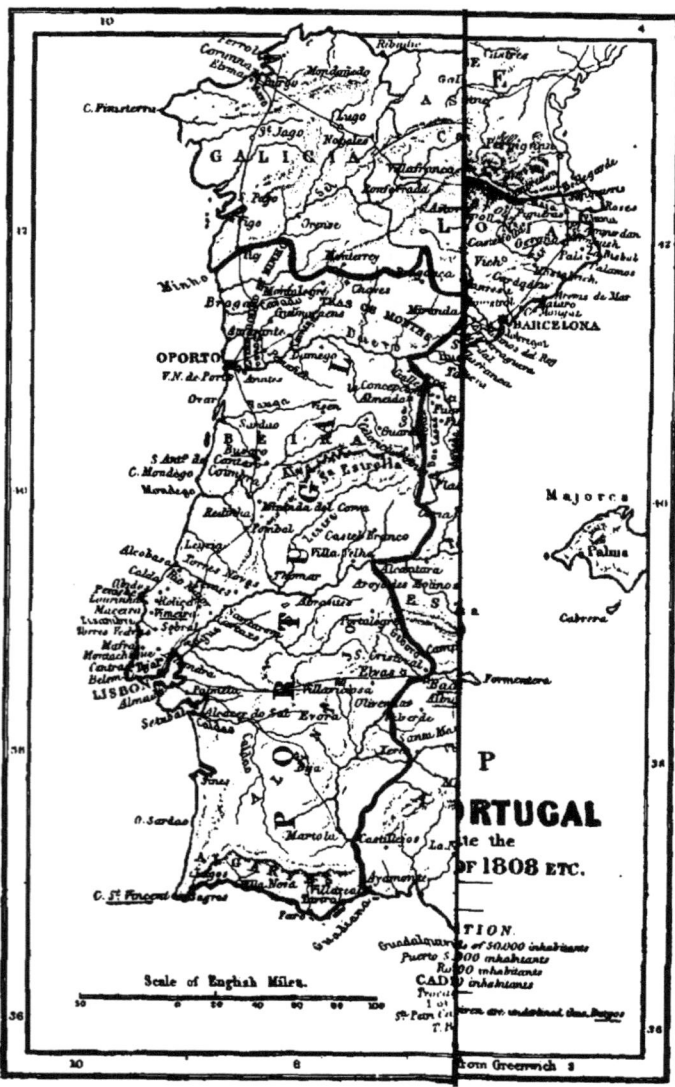

# CHAPTER IV

1808 to 1809—Portugal—Roleia—Vimiera—Spain invaded—Talavera.

Soon after signing the Peace of Tilsit, Napoleon called upon Portugal to shut her ports against England, to confiscate English property, and detain all Englishmen found within her borders. The Government of Portugal temporised, but ultimately surrendered; whereupon Lord Strangford, the British Minister, demanded his passports, and went on board the British fleet, lying in the Tagus. Not satisfied with this, however, Napoleon declared that "the House of Bragança had ceased to reign," and sent an army under Junot to carry out his decree; in consequence of which, on the 29th of November, the Prince Regent retired to Brazil with the whole Portuguese fleet of thirty-six sail, and escorted by four British line-of-battle ships.

Within a few days the heads of Junot's columns appeared on the hills above Lisbon, and the port was evacuated by the British.

In March, 1808, the French occupied Madrid, and in May the populace of the city rose *en masse* and massacred their temporary masters. The province of Asturias followed suit, and created an independent Junta; and their example was followed in every province of Spain not actually held by the French. These Juntas sent delegates to London, who arrived there on the 9th of June, and demanded the interference of Great Britain in the Peninsula.

Admiral Sir Charles Cotton, who at this time commanded on the coast of Portugal, so far exerted himself in the

Portuguese cause that, when Oporto rose against the French, and drove them south of Madrid, they were followed by the whole of the north of Portugal; thus making an opening which was immediately taken advantage of.

The English Government acted decisively, and an army of about 10,000 men was assembled at Cork, among which was included the 45th, still under Lieut.-Colonel Guard, with Major Gwyn as second in command. The whole force was under Sir Arthur Wellesley, who had recently won a great reputation in India, and was at this time holding the post of Chief Secretary for Ireland.

The force embarked on the 12th of July, 1808, with 346 cavalry and 12 guns, but absolutely deficient of commissariat or medical establishment.

"It certainly was a shabby enough start," said Sir Arthur Wellesley in after years, "but it was quite of a piece with our military policy of the time. The Government trusted me, I believe, as much as it trusted anybody, but it had no faith even in me as yet, and dreaded nothing so much as throwing a large army ashore on the Continent under the command of a British officer. I must admit, however, that the men were admirable, and admirably drilled. All that they wanted was experience, and that they got by degrees."

The embarkation strength of the 45th was 35 officers, 38 sergeants, 37 corporals, 22 drummers, and 635 privates.

The expedition, after some delay off Finisterre, while Sir Arthur Wellesley was conferring with the Juntas, finally landed at Mondego Bay, about ninety miles north of Lisbon.

The object in landing was to turn the French out of Lisbon, and the co-operation of the Portuguese troops raised in the north, and already under British officers, was of paramount necessity, which was no doubt the primary reason for choosing a spot so far from the actual objective; moreover, the village of Leyria, not far south of Mondego, was held by the French. The Portuguese army was assembled at Coimbra, just inland to the north of Mondego, so it was obvious that the landing would be unopposed, and that Sir Arthur Wellesley would have leisure to arrange his forces before marching direct upon Lisbon.

The disembarkation, however, had hardly commenced when Sir Arthur Wellesley received despatches from England informing him that all the plans were altered, that a much larger force was to join the army, and that he was to be superseded in the command by Lieut.-General Sir Hew Dalrymple from Gibraltar. One brigade was following under Anstruther, another under Ackland, while the army, under Sir John Moore, which had been moved up the Baltic, was on its way out with Sir John Moore himself, who, in his turn, was to be superseded by Sir Harry Burrard.

The disembarkation, including that of General Spencer's corps, was completed by the 6th of August, and the 45th, 50th, and 91st were formed into the 5th brigade, under the command of Brigadier-General Catlin Crauford.

The French at this time had an advanced post under General Laborde at Roleia, fifty or sixty miles south of Mondego; and a corps under General Loisson was marching from Elvas to cross the Tagus at Abrantes and support it. Consequently, Sir Arthur Wellesley determined to march at once and crush Laborde before he was reinforced by Loisson, without waiting for the changed policy of the home Government to take effect.

Having scarcely any commissariat and no train, there could be no base at Mondego, consequently he was compelled to march with his right flank in touch with the fleet, and to depend upon it for supplies. Five thousand of the Portuguese troops were selected to advance with the army, but only some 1400 infantry and 250 cavalry actually co-operated,

D

the remainder holding to the rear owing to the difficulty of supply from that direction by land.

On the 10th of August the advance commenced, the mounted Dragoons leading, followed by the 3rd, 5th, and 4th brigades of infantry; and the same evening the force halted at Lugar, the advanced piquets being furnished by the 5th brigade.

Leyria was reached on the 12th, and on the 15th the French advanced guard at Obidos was driven in by two companies of the 95th Rifles and two companies of the 60th. On the following day the army halted before Roleia.

The French troops were posted at the foot of the high ground on which Roleia stood, with a broad plain on their front and flanks stretching away to the hills which formed the valley at the south end of which the town was built. Away to the south, in their rear, the ground became more mountainous, with paths and passes leading up to a stretch of table-land.

Sir Arthur Wellesley had 13,700 infantry, 650 cavalry, and 12 guns; with so vast a superiority there could of course be no doubt as to the result, but in any case Laborde's object was not an obstinate defence. He wished only to check the British advance sufficiently to allow Loisson to join him, so that the two together could show a strong front.

Coolly aware of the whole situation, Wellesley sent out a light column over the hills on either flank to turn the French, while his main body marched straight to its front upon their centre.

The order to advance was given at seven o'clock, Crauford's brigade supporting the artillery along the high road leading through the centre of the enemy's position; the honour of driving the French from their ground falling to the lot of the 45th and the light companies. The enemy retired through the passes up the mountains, closely followed by our infantry,

—the ground being unfitted for cavalry—the 45th and 82nd scrambling up the very rough ground on the extreme left. In places the enemy stood, defending the ground hotly, but nothing could stop the advance of our troops, who eventually reached the table-land to find the French in full retreat, having left behind three guns, and over 1000 men in killed, wounded, and missing. The British bivouacked on the ground they had taken after this the first of our Peninsula victories.

On the conduct of our men Sir Arthur Wellesley said:—
"I cannot sufficiently applaud the conduct of the troops throughout the action. The enemy's positions were formidable, and he took them up with his usual ability and celerity, and defended them most gallantly."

The 45th lost Ensign Dawson, who carried the King's colours, killed, Lieutenant Burke and 9 rank and file wounded; while the staff of the regimental colours was cut in two by a cannon shot. The steep ascent climbed during the pursuit of the French proved fatal to Captain Payne, who had been shot through the lungs at Buenos Ayres, and whose wound was reopened by the exertions he put forward; he had to fall back to the rear, and did not long survive.

Had he considered himself free, Sir Arthur Wellesley would probably have marched on Torres Vedras at once, but requiring supplies and hearing that General Anstruther was close off the coast with a fresh brigade, he marched his army to Lourinhal, close to the sea, on the 18th, and there received the supplies he required.

On the 20th of August Sir Harry Burrard, who himself was expecting to be shortly superseded by Sir Hew Dalrymple, arrived in Marceira Bay, and Sir Arthur Wellesley at once went on board to report, and hand over his command.

By the arrival of Ackland's and Anstruther's brigades

there were now more than 17,000 men in Portugal, and Sir
Arthur Wellesley urged upon the new commander that the
proper course to pursue was to march at once round Torres
Vedras to Mafra, while Sir John Moore, landing his troops
at Mondego Bay, should move directly inland upon Santarem; this would have turned the enemy's position in front
of Lisbon, and cut off his retreat across the Tagus upon Elvas.

Sir Harry Burrard, however, for some inexplicable reason,
refused to listen to these proposals, and instead sent to Sir
John Moore to suspend his landing, and bring his troops
down by sea to Marceira Bay.    He then sent Sir Arthur
Wellesley on shore again, and remained on board himself,
where it was impossible he could direct the movements of·
the army he had taken out of its commander's control.   Sir
Arthur Wellesley accordingly came on shore again, and
prepared for the attack which he knew the enemy was sure
to make.   His troops were fairly well disposed, lying in line
on a height, with the right flank resting on the sea, and the
5th brigade and the Portuguese in reserve in rear.    He
calculated that Junot had not control of more than 14,000
men, and that he must have some 3000 of them to garrison
the forts at Lisbon.

At nine o'clock the next morning the expected attack
came, the French advancing in three columns.   The ground
immediately in front of the British position was so wooded
that the intentions of the enemy could not at first be clearly
made out, but it soon became apparent that the main attack
would be developed on the British left.   This was, therefore,
extended and strengthened, the whole line facing Torres
Vedras, with the centre advanced and the flanks somewhat
thrown back; the 5th brigade and the Portuguese still being
kept in reserve.

The French attack was somewhat on all fours with
Wellesley's attack at Roleia; a turning column on the left

BATTLE
OF
VIMEIRA.
21st August 1808.

Cavalry   Infantry   Artillery
                      (Allies)
                      (French)

SCALE
Military Scale 2¼ Feet each
English Miles

ATLANTIC OCEAN

attacked the British right under Anstruther, another turning column on the right under Brennier endeavouring to turn the British left commanded by Ferguson, while the reserve and cavalry under Kellerman attacked the British centre. The execution, however, was defective; the French right was, for some reason, delayed, and did not get into action until their left attack had been completely routed by the superiority of the British fire, backed by the stern threat of the bayonet. The French right was driven back in half-an-hour, and all Kellerman's charges in the centre were equally futile. The field was easily won, and Sir Arthur Wellesley ordered the whole army to advance in pursuit. At this point, however, Sir Harry Burrard, who had landed during the height of the battle, interfered, and, stopping the pursuit, ordered the British to remain on the ground they had held. The 45th, who had been in reserve all day, and suffered no casualties, were thus deprived of that share in the glories of Vimiera which their active courage merited; but their steadiness in reserve and alacrity in movement did not escape the notice of the chief, who specially complimented them. before he left the army for England.

The day after the battle Sir Hew Dalrymple arrived to take the supreme command of the army, and Sir Arthur Wellesley, having a rooted objection to serving under him, left for home to resume his duties as Chief Secretary for Ireland, his departure being hastened by the news of the death of his deputy.

Previous to his departure, however, on the 23rd, Junot had sent Kellerman with a flag of truce to propose a suspension of hostilities as a preliminary to capitulation. It was probably clear to every one that Junot's condition was desperate, and that any terms might have been imposed upon him; but Dalrymple, apparently overjoyed with a measure of success in no way due to his own acts, allowed Kellerman

to draw up the terms of the truce, which he ordered Sir
Arthur Wellesley to sign, and subsequently carried through
the disgraceful Convention of Cintra, which Wellesley's sense
of military honour compelled him to bear the blame of for
months, until the official enquiry showed him in his right
light.

The greater part of the enemy moved into quarters in
Lisbon; and all the field officers subscribed to present Sir
Arthur Wellesley with a piece of plate in remembrance of
the first successful campaign against the French.

After the battle of Vimiera the 45th took up their quarters
at Torres Vedras, whence Captain Greenwell was despatched
with a party to the fortress of Peniche, to summon the
French garrison there to surrender; this being done, the
succeeding garrison was supplied by the 45th.

Dalrymple and Burrard were soon afterwards relegated to
that obscurity from which they should never have been
allowed to emerge, and Sir John Moore became commander-
in-chief of the army in the Peninsula.   With his sad but
glorious story the 45th had no connection, as when the army
went north into Spain they remained in garrison at Lisbon.

As a result of the disastrous consequences of removing the
base of operations from Portugal, which culminated at
Corunna, the Government, almost in despair, turned again
to Sir Arthur Wellesley, who accordingly returned to the
Peninsula, and landed at Lisbon on the 22nd of April, 1809.
Here he found that during the previous month Marshal Soult
had carried Oporto by storm against the Portuguese, and was
now occupying the city with 20,000 men; while Marshal
Victor, with 28,000 men, was at Merida, in Estremadura.
At the disposal of the British commander were some 35,000
men, including German and Portuguese troops, of which he
assembled some 13,000 English, 9000 Portuguese, and 3000
Germans at Coimbra, for the purpose of striking at Soult.

At the same time it was necessary to guard against an advance on Lisbon by Victor; for this purpose the defensible passes were occupied, and the bridges over the Tagus broken down. The 45th were brigaded with the 27th and 31st, under Major-General Mackenzie, and the whole brigade occupied Santarem, Abrantes, Villavellia, and Alcantara, along the line of the Tagus, the 45th being at Villavellia.

Marshal Soult having been driven out of Oporto with a suddenness and completeness which reduced his army to a mere wandering crowd among the mountains of Northern Spain, Sir Arthur Wellesley came down south with the object of co-operating with the Spanish army under Cuesta.

In spite of their hitherto unchecked career of victory, the British army was admittedly facing tremendous odds; the French had 175,000 men and 33,000 horses in Spain, while Sir Arthur Wellesley could not put into the field more than 21,000 British and Germans, together with 17,000 Portuguese, over whom he can hardly be said to have had complete control; while the Spanish army of some 115,000 men was not altogether to be relied on.

The French were massed in four great bodies. King Joseph was covering Madrid with 58,000 men; his most advanced troops under Marshal Victor, occupied the valley of the Tagus about Placencia. To the north, concentrating about Salamanca, was Marshal Soult, with 54,000 men. Far away to the north-east, in Aragon, were 18,000 men under General Suchet; and still further off in Catalonia, Angereau had concentrated some 31,000 men.

After conference with Cuesta, Wellesley determined, in conjunction with the 38,000 Spanish troops in Estremadura, to advance up the valley of the Tagus by Placencia, Oropesa, and Talavera, drive back Victor, and strike at Madrid. Probably, when he came to this determination he was ignorant of the capacities of the Spanish troops, and of the

fact that Soult was in such strength at Salamanca.    He,
moreover, fully believed that while he was marching up the
valley of the Tagus with Cuesta, 25,000 Spaniards, under
Venegas, would make a flank attack from the south, crossing
the Tagus at Toledo; and that the Spanish force in Galicia,
under Romano, together with the Portuguese operating in
the neighbourhood of Ciudad Rodrigo, would be able to hold
the northern French army in check.

Sir Arthur Wellesley marched into Spain on the 27th of
June at the head of 21,000 men, but soon discovered the
value of Spanish promises; nothing turned out as had been
intended, and the promised supplies and transport utterly
failed.   The movement, however, alarmed Victor, who fell
back upon Talavera, behind the Albreche, and Wellesley
occupied Placencia on the 16th of July.   He again advanced
on the 18th of July, the troops carrying provisions to last
until the 21st, and on the 20th he was at Oropesa in personal
communication with Cuesta.   Here, for the first time, he
saw the Spanish army, and found it to be nothing but a dis-
organised rabble.   However, there was nothing for it but to
go on, and he considered that the mountains on the left of
the line of advance were a sure protection from attack on
that flank.

Before leaving Placencia the army had been reorganised,
and the 45th, with the 24th and 31st regiments, formed the
1st brigade of the third division under the command of
Major-General Mackenzie; the division formed the advance
guard of the army, and led the advance towards Talavera.

It was understood that Victor was in some force at Talavera, and on the 23rd, dispositions were made for attack next
morning.   Mackenzie's division, however, advancing on the
24th, found the place evacuated, and, crossing the Albreche,
proceeded some distance on the road to Madrid without
meeting the enemy, and there bivouacked.

Cuesta, who was well supplied, and believed himself quite capable of dealing with the retreating French, pushed on when Wellesley halted on the right bank of the Albreche, and meeting the whole of the forces under King Joseph's command, was completely routed and driven back in disorder. Wellesley arrested their retreat by mere force of character, assumed the command-in-chief, and supposing his rear safe, placed his army in the defensible position which the neighbourhood of Talavera afforded.

The town of Talavera is situated close to the north bank of the Tagus; to the northward the ground rises for a couple of miles until it reaches a point of considerable height, dipping to a valley, which is commanded by a similar height a gunshot's distance off. Entrenchments and two field redoubts had been thrown up in front of Talavera, extending from the river some three-quarters of a mile towards the height to the northward; while in front of the position ran a water-course, beyond which dense olive woods extended for some two or three miles.

The Spaniards, some 33,000 strong, were placed in the town of Talavera, and behind the entrenchments in front of it. The British—19,000 men and 30 guns—were extended towards the height on the left, but, oddly enough, not at first occupying the highest points. The slope towards the front was steep and difficult, that to the rear gentle and easy, which gave much local strength to the British posts. The weak point in the line was the British extreme left, where the ground dipped into a rugged valley, to rise into the mountain chain beyond; but even here the water had worn a chasm which was a hindrance to the attack.

On the evening of the 27th the line of defence was not completely taken up. Mackenzie's division and a brigade of cavalry were still thrown among the woods, a couple of miles in front, in the neighbourhood of a large house, the Casa

Salinas, to the roof of which Sir A. Wellesley and his staff had ascended in the afternoon in order to gain the earliest view of the enemy's approach.

About three o'clock the British outposts were surprised by Lapisse and Ruffin's divisions, and the Commander-in-Chief and his staff, who were on the roof of the Casa, barely escaped capture. The English brigades got separated, and were driven in disorder from the forest, a disaster being only averted by the steady conduct of the 45th and some companies of the 60th Rifles.

There is no doubt that the conduct of the regiment on this occasion was magnificent. The Commander-in-Chief himself in his despatch says: — "He had opportunities of noticing the gallantry and discipline of the 5th battalion, 60th and the 45th, on the 27th"; while Lieut.-Colonel James Campbell, who was present on the occasion, writes in *A British Army*: —"Some may remember that on the 27th of July, 1809, the first day, I may say, of the battle of Talavera, the enemy's light troops broke unexpectedly in upon us at the Albreche river, when our troops were quite unprepared for such an event; some young corps were surprised, and consequently did not behave well. Lord Wellesley himself, if I mistake not, and some of his staff were placed in a very perilous position in an old house, into which they had gone in order to ascertain, from its highest window, what was going forward, and his lordship had immediately after to take upon himself, in a great measure, the direction of the hastily-formed rear-guard of infantry, consisting of the 45th regiment and 5th battalion 60th, which corps, assisted by our cavalry, covered the retreat of the advanced division till it reached the position in which the battles of that night and the following day were fought." The retirement, which, as was afterwards ascertained, excited the admiration of the French general, was made by wings

and companies so far as the nature of the ground permitted;
and Colonel Guard being wounded, the command passed into
the hands of Major Gwyn. It became necessary to separate
the regiment into two parts; six companies under the Major
retired upon the centre of the allied line, while four companies under Captain Smith, covering the retreat of some
other portion of the forces, fell back upon the Spanish lines.

The French cavalry were close upon them, and just as the
four companies got among the Spanish army it broke and
fled in the dusk, sweeping them away with it. Though some
of them subsequently reformed and returned to their posts,
the right of the allied army was 6000 short of its numbers
next day, and one of the redoubts was silent for want of guns.
Captain Smith and his officers used all their endeavours to
extricate their men, and succeeded in bringing the four
companies up in rear of the extreme left by daylight next
morning.

Victor, on the French right, observing the weak point of
the English position opposite to him, made a vigorous dash
to turn the position and occupy the hill as twilight was
deepening into night. His men actually gained the summit,
but were driven back, and the English secured their posts
for the night.

Soon after daylight the French renewed the attack,
making, as before, desperate efforts to turn the English left
and carry the heights there. The fighting was desperate, but
in forty minutes the French were driven back with a loss of
1500 men. To secure his left, Sir Arthur Wellesley obtained
from the Spaniards a division under Bassecour, which he
placed on the slope of the mountains beyond the valley for
the purpose.

The four companies of the 45th, under Captain Smith,
took part in this gallant defence under Hill, and were
in the thick of it the whole morning, after which they

rejoined the rest of the regiment near the centre of the English line.

Napier tells us that " from nine o'clock in the morning until mid-day the field of battle offered no appearance of hostility. The weather was intensely hot, and the troops on both sides mingled, without fear or suspicion, to quench their thirst at the little brook which divided the positions. Before one o'clock, however, the French soldiers were seen to gather round their eagles, and the rolling of drums was heard along the whole line."

The whole French army now advanced with vigour against all the English positions. One French column actually worked past the left of the 45th at a distance of not more than fifty paces. Captain Greenwell, commanding the left wing, at once threw back three companies, who poured a heavy fire into the mass of the French, who were at once checked and finally driven back down the slope in disorder. At the same time a direct attack was made on the right wing of the regiment, which was repulsed with equal determination, but not until heavy loss had been inflicted on the gallant "stubborns."

Beyond the 45th the line was broken by the French, but was restored by the 48th; while the attack on the valley to the extreme left was rendered abortive by Wellesley's care and the cavalry; and the hill which could not be taken when half held, became impregnable when held in force. The repulse of the French from the centre was the note of triumph all along the British line; the enemy fell back beaten, and our victorious and hungry troops rested on the position they had so fiercely held.

But the victory had cost us dearly; 6200 were returned as killed, wounded, or missing; while the French loss was nearly 7500. In the two days' fighting the 45th loss was:—
13 rank and file killed; 5 officers: Lieut.-Colonel Guard,

Major Gwyn, Lieutenants Robinson and Cole, and Ensign
Milne; 4 sergeants and 130 rank and file wounded; and
Brevet-Major Lecky, 1 sergeant, and 10 rank and file
missing; in all 182.

The following story is told of Private Mansfield, who had
served with the regiment in the West Indies prior to its
return to England in 1801. He was rather an awkward man,
yet a good soldier withal, though never, perhaps, in the
strictest sense of the word, a smart one; he served in the
light company, and was conspicuous for his bravery during
the attempt on Buenos Ayres. He was mortally wounded at
Talavera, and when after the action they sought to move
him, he refused help, but begged to see his Captain—Captain
Greenwell. This officer went to him at once, and remon-
strated with him on his resolution.

"It's of no use, sir," said the dying soldier; "my hour
is come; but I have two requests to make which I rely on
you to answer justly. Have I done my duty as a good and
a brave soldier?"

"Most undoubtedly, Mansfield," returned the captain;
"like an excellent one."

"The second is, that you will promise me that the balance
of my accounts shall be transmitted to my wife."

This being also promised, he breathed his last, saying—
"Well, then, I die happy and contented."

Two days after the battle, Soult drove away the insig-
nificant guard that Cuesta had posted at the pass of Banos,
and on the 31st of July occupied Placencia at the head of
53,000 men. Joseph at the same time was on the far side
of the Albreche with 40,000 men; consequently Wellesley,
who had arranged with Cuesta to hold Talavera, in order to
protect the hospitals and wounded, and who had not more
than 23,000 troops at his disposal, was in an apparent trap,
from which nothing but the most consummate skill could

extricate him.   The only line of retreat was across the Tagus, by the three bridges at Talavera, Arzobispo, and Almarez.

On the 3rd of August the English army was at Oropesa, and at five o'clock in the evening Wellesley learnt that the French were between him and the bridge at Almarez; an hour later news arrived from Cuesta that, in the presence of an advance by Joseph, he had abandoned Talavera, leaving the wounded prisoners in the hands of the French.   Lieut.-Colonel Guard and Lieutenant Cole thus became prisoners of war.

With the bridges at Talavera and Almarez in the hands of the French, the only course left open was to move the whole army across the bridge at Arzobispo, and march back into Portugal by the south bank of the Tagus.   For this purpose Wellesley moved off on the following day, and by two o'clock in the afternoon the whole force was safely across the Tagus; a defensive position being taken up on the left bank. Soult and Ney, however, deemed it expedient to move northward, while Victor and King Joseph withdrew to the eastward to check the Spanish in La Mancha, thus leaving Wellesley free to act; but deciding on the impracticability of resuming the offensive, he withdrew his army on the 20th, and marching by Truxillo and Merida, established his headquarters at Badajoz on the 4th of September.

Thus ended the two months' campaign, in which the 45th had been concerned from first to last.   A brilliant victory, it is true, had been gained, but no results had followed; the losses had been considerable, and a bitter hatred of the Spaniards had been engendered among our overworked and ill-fed soldiers; while, to crown all, the pestilent fever of the Guadiana and dysentery had made terrible ravages among the troops.

So indifferent were the commissariat arrangements that it is related that when crossing the bridge at Arzobispo the army fell in with a herd of swine, the pangs of hunger were so great that the men broke the ranks and fell upon the animals like wild beasts, even cutting the flesh off their bodies while they were still alive.

But in spite of being a failure, the campaign had clearly proved two things—the steadfastness and ability under great trials of the English general and the stubbornness and capacity for fighting, under the most adverse circumstances, of the troops he commanded.

THE news of the Talavera campaign was received with enthusiasm in England ; the thanks of both Houses of Parliament were voted, and Sir Arthur Wellesley was created Baron Douro and Viscount Wellington.  The King caused the Commander-in-Chief and Secretary of State to write letters of thanks to the army, in which the following passages occurred : —

"The King, in contemplating so glorious a display of the valour and powers of his troops, has been graciously pleased to command that his royal approbation of the conduct of the army serving under Lieut.-General Sir Arthur Wellesley shall be publicly declared in general orders.  .  .  .

"It is His Majesty's command that his approbation and thanks shall be given in the most distinct and most particular manner to the non-commissioned officers and private men. In no instance have they displayed with greater lustre their native valour and characteristic energy ; nor have they on any former occasion more decidedly proved their superiority over the inveterate enemy of their country."

Nor were the general public less forward in their recognition of the splendid work that had been done.  Captain Drew, who had taken home the invalids of the 45th, marched with 76 of them through Leicester, on his way to Nottingham, where the second battalion was quartered ; the people there subscribed 3s. 6d. per head for the men of Talavera.  Captain

Drew, however, with nice discrimination, pointed out that only 49 of the 76 had been actually engaged in the campaign, and made a present of money subscribed for these 27 to the Female Hospital of the town.

After the Talavera campaign the British, as we have already seen, were lying at Badajoz, Elvas, and neighbouring places. The French at the same time were in full possession of the north coast of Spain from San Sebastian to Gijon, as well as nearly the whole of the central part of the Peninsula; while Portugal, Galicia, Andalusia, and parts of Estremadura, Murcia, Valencia, and Catalonia were under the control of the allies; but in many parts, especially in Catalonia, that control hung upon the life of a few fortresses, such as Gerona and Lerida.

The French, while considerably outnumbering the allies, had the great advantage of occupying the interior position, and were continually recruited and well supplied from France. They were thus far better provided, while their discipline and order were beyond comparison with all but the British portion of the allied forces.

Lord Wellington, after carefully reviewing the situation and the conditions and capabilities of the opposing forces, was convinced that if the autumn and winter of 1809, and the spring of 1810, were spent in an attitude of stubborn defence, with an absolute renouncement of any temptation to advance in any direction, the Spanish and Portuguese armies might be so organised and disciplined as to be capable of much useful work during summer. It was also apparent to him that, apart from the invasion of Portugal, the French armies could do little so long as the Spanish forces stood on the defensive. Moreover, the invasion of Portugal offered by no means an easy task.

There were only three practicable routes for the march of the French armies from Spain into Portugal. That from the

z

north, from Salamanca, through Ciudad Rodrigo across the
frontier to Almeida and Guarda, and on to Lisbon by the
valley of the Mondego.    The middle route was by the
valley of the Tagus, through Placencia and into Portugal by
the route Wellington had followed into Spain in June and
July.    The third, the most southerly, was by Badajoz, across
the frontier to Elvas, leading to Lisbon by crossing the Tagus
at Abrantes.

The southern and northern routes were both guarded by
fortresses, which required to be taken, or thoroughly masked,
before an advance could be made in safety by either of these
roads.    The northern road was covered by Ciudad Rodrigo
and Almeida, and the southern by Badajoz and Elvas.

In the autumn of 1809 the allied armies were roughly
disposed as follows:—There was a Spanish army in Galicia
under the Duke del Parque; Beresford's Portuguese army
was covering Ciudad Rodrigo and the northern route into
Portugal; there was a Spanish army of, perhaps, 50,000 men
in Estremadura and Andalusia covering Seville and Cadiz;
while Lord Wellington, as we have already seen, was at
Badajoz with the British army, covering the southern route
into Portugal.

Late in September the Duke del Parque demanded the
assistance of the Portuguese army to enable him to advance
against the French from Ciudad Rodrigo; this being refused,
he proceeded alone, gained a small and undecisive victory
over the French general Marchand, at Tamames, pushed on
past Salamanca, and then fell back to Alba de Tormes.    Here
the approach of the French, under Kellerman, caused the
whole Spanish army to throw away its arms and disperse
among the mountains.

A similar but rather heavier disaster befell the southern
Spanish army, which had advanced rapidly to the north, and
was attacked by the French, under Soult and Victor, at

Ocaña on the 20th of November. The result was the practical destruction of the Spanish army, which lost over 30,000 in killed, wounded, and prisoners, with all its baggage and some 45 guns.

The French, in consequence, under King Joseph, quickly overran Andalusia, making themselves masters of the whole province, except Cadiz, which, being open to the sea—though blockaded on the land side—held out by the aid of English stores, men, and money.

The dispersal of the Spanish army under del Parque, combined with Wellington's belief that the French would invade Portugal from the neighbourhood of Ciudad Rodrigo, and the fatal spread of sickness in his camp, determined him to transport his army to the north, and before the end of the year he had established his headquarters at Viseu, and disposed his forces in the valley of the Mondego. He had, however, left General Hill with 10,000 British and Portuguese troops at Abrantes to succour Elvas and Badajoz if necessary, and to cover the approaches to Lisbon by the north bank of the Tagus.

Early in 1810 Major-General Picton joined the army, which, on the 22nd February, was reorganised in five divisions of infantry and one of cavalry.

The first division was commanded by Spencer, the second by Hill, the third by Picton, the fourth by Cole, and the light division by Robert Crauford, an old 45th man of whom we have already heard at the attack on Buenos Ayres. The 45th, brigaded with the 88th and 74th regiments, formed the second brigade of the third division, which soon became well known under the sobriquet of the "Fighting Third."

By March it had become clear that the whole of the French forces were in motion towards Portugal. To oppose this army of some 200,000 men the British had only about 23,000 men in the field, supported by some 30,000 Portuguese, while possibly another 20,000 Portuguese were disposed to the north and south.

Wellington, whose plan was to hinder the French advance by every means in his power—and if this was found impossible, to devastate the country behind him, and fall back on the lines of Torres Vedras—fell back from Abrantes and Viseu to a central position, ready to operate on whichever of the three routes the French might adopt. The main attack ultimately came, as Wellington had expected, by Ciudad Rodrigo, a French column of 97,000, supported by another corps of 11,000, advancing by that route.

Ciudad Rodrigo is situated on the Agueda River, which, running in a northerly direction, joins the Douro. Fifteen miles to the west of the Agueda is the Coa, which also, running in a northerly direction, joins the Douro. The road from Ciudad Rodrigo, running at first to the westward, crosses the Coa close to the fortress of Almeida, which is near the east bank of that river. On the east of the Coa is a broken mountain chain, traversed by defensible passes, which follows the course of the river southward for some 60 miles, when it turns to the south-west, and runs down for about 130 miles towards Lisbon, where it is lost in the valley opposite the lines of Torres Vedras. From Almeida the road leads directly over the hills in a south-westerly direction into the valley of the Mondego; while a branch road from the Coa bridge leads more to the southward to Guarda, which commands another entrance into the Mondego valley.

From his central position in the valley Wellington sent the light division and some cavalry, under Crauford, to operate between the Agueda and the Coa, with the fourth division, under Cole, in support at Guarda; while Picton's division, in which were the 45th, was advanced to Pinhel. Crauford, though it was clear he must eventually be driven

back, maintained his ground cleverly, and it was not until
the 25th of June that he began to retire on the Coa.  On the
25th of July, after a most brilliant rear-guard action on the
Coa, he was pressed back across that river by Masséna, who
thus secured the passage of the river and the choice of his
route for the advance upon Lisbon.

In consequence of the uncertainty of the direction from
which Portugal would be invaded, Wellington had been
obliged to keep Hill's division, together with Leith's troops,
covering Lisbon to the east and south-east; this weakening
of his forces had made it impossible for him to proceed to the
relief of Ciùdad Rodrigo, which fell on the 11th of July;
and Almeida, which was closely invested by the French, after
Crauford's retirement, also capitulated on the 27th of
August; Masséna, whose movements were the reverse of
rapid, made no advance until the 6th of September, when
Reynier's corps was ordered to concentrate at Guarda; Ney's
corps and the heavy cavalry at Macal de Cheo, and Junot's
corps at Pinhel.

On the 12th the British were driven out of Guarda, on the
13th they reoccupied it, but were finally driven out on the
15th.   Reynier's corps then advanced into the valley of the
Mondego, and, effecting a junction with Ney and the cavalry,
crossed to the right bank of the river.   Wellington, satisfied
as to the nature of the movement, sent orders to Hill and
Leith to join him, and, gradually falling back before
Masséna, took up his position on the heights of Busaco, and
there offered battle to the French. . The position was a
highly-defensible one, dominating the five roads leading out
of the valley on the north bank of the Mondego.   One road
ran round the mountains to the north of a hill called Santa
Caramula, which was not occupied; another led to the south
of the range close to the river, while three roads in the centre
led over the mountain itself.   A deep gorge, the sides of

which were in places thickly wooded, ran in front of the British position, and separated the two armies. .

On the 25th of September, while Masséna himself was still ten miles in rear, Ney approached the position and saw a great opportunity. The British were neither in full force nor in full position; Spencer's division was not up, Leith was only crossing the Mondego, while Hill's division was still further off on the other side of the river. Crauford's light division was in touch with the enemy at the bottom of the gorge, and the remainder were only moving up the side of the hill into their assigned positions. Ney pressed for leave to attack, but Masséna, who could not be present, would not allow it, and the British thus gained the twenty-four hours' respite which they required; Masséna not arriving in view of the position till noon on the 26th, by which time Wellington's troops were all formed up.

Hill was placed on the south, guarding the road and the fords over the Mondego; Leith occupied the next hill, having the road from St. Antonio da Cantara on his left; some two miles on his left, on the further side of a depression, came Picton's division, whose left was separated by nearly a mile from the right of the first division; the light division was to the front and somewhat to the left of the first division, while Cole's division occupied the extreme left of the position.

The St. Antonio road was the lowest point in the position, and offered the easiest approach to the enemy; the 74th were placed on the road, supported on the right by two Portuguese guns, and still further to the right by the 9th and 21st Portuguese regiments; on the immediate left of the 74th were the 8th Portuguese, and then came the 45th and 88th regiments.

Soon after dawn on the 27th, while the mist still hung about the valley, the enemy's attack began to develop. Three heavy columns advanced against the left of the allied army and two against the third division. Picton, observing the gap between his division and Spencer's, moved the 88th, supported by four companies of the 45th under Major Gwyn, further to the left; five companies of the regiment remained in position on the right under Lieut.-Colonel Meade, who had recently joined from the second battalion.

The portion of the French column which came up the St. Antonio road was easily dealt with by the 74th and its supports, who drove them back with small loss to themselves. Picton seeing this, and hearing increased firing from the left, rode away in that direction, leaving orders for Colonel Meade to move at once to the left with his five companies and the 8th Portuguese. The French, by the energy of their attack, had, however, gained the higher ground, and Picton, before he had got half way, found the light companies of the 74th and 88th retiring in disorder, and the 8th Portuguese broken and in confusion. At this critical moment Major Smith of the 45th rode up, and with Picton succeeded in rallying the yielding troops; Major Smith placing himself at the head of the light troops gallantly charged the head of the French column, and drove them down the hill again; but the brave leader laid down his life in attaining his object.

During these events a far fiercer encounter was being waged on the left; there, the French column was endeavouring to force its way between the 88th and the four companies of the 45th; in overwhelming numbers the enemy came on, and the position was critical, but Colonel Wallace, of the 88th, was capable of dealing with it. Waiting until the French were close up he gave the order to charge, and the Connaught Rangers and the 45th went madly at them with a ringing cheer. No troops could have withstood them, and in one mingled mass, attacked and attackers went down the mountain side, their track marked with dead and dying, to the bottom of the valley; and the right column of the French left attack was beaten and destroyed.

A large body of French who had passed round the left flank of the 74th were effectually dealt with by Leith, while Ney's attack on the left centre of the allied position was completely and decisively repulsed by the 43rd and 52nd regiments; and the whole French army were driven down to the bottom of the gorge, where the reserves covered their retirement, and by two o'clock a spontaneous truce had set in, and both sides were employed in carrying off their wounded.

In this decisive encounter the enemy lost about 4500 men, while the loss of the allies did not exceed 1300. Of the 45th, Major Smith, Captain Urquhart, Lieutenant Ouseley, one sergeant, and 21 rank and file were killed; while Major Gwyn, Lieutenants Harris, Tyler, and Anderson, 3 sergeants, and 106 rank and file were wounded. These losses were entirely in the four companies led by Major Gwyn, whose conduct cannot be too highly extolled. Conquest, after a loss of 42 per cent., tells its own story.

Picton, in writing to the commander of the forces, said :— " Your lordship was pleased to mention me as directing the gallant charge of the 45th and 88th regiments, but I can claim no merit in the executive part of the brilliant exploit which your lordship has so highly and so justly extolled. Lieut.-Colonel Wallace and Major Gwyn, who commanded the four companies of the 45th engaged on the occasion, are entitled to the whole of the merit; I am not disposed to deprive them of any part."

Lord Wellington, in his public despatch, said :—" I beg to assure your lordship that I have never witnessed a more gallant attack than that made by the 88th, 45th, and 8th Portuguese regiments on the enemy's division, which had reached the edge of the sierra." It is, moreover, related that Lord Wellington, ever sparing of praise, could not contain himself on witnessing the charge. "There, Beresford! Look at them now!" The exclamation, spontaneously forced from the lips of the chief, was something like a triumphant vindication of the opinion he had already formed of the regiment at Talavera. The "stubborn old regiment" had shown itself more than stubborn; it could stand and endure; it could also rush headlong upon forces overwhelmingly superior and defeat them.

Colonel Meade, who commanded at Busaco, had only recently joined from the second battalion; he returned to England before the British forces quitted the lines of Torres Vedras. Possibly there were reasons for his hasty retirement, if the following story, taken from Robinson's "Life of Picton," is any indication of his general character:—

"The night was cold, and the position occupied by the troops exposed them to the inclement blast which swept over the mountains; even the hardy veteran shrank within his scanty covering. The young soldiers, however, and even the young officers, endured with much less patience their mountain couch. A party of these latter (to one of whom we are indebted for this anecdote) tired of the coldness of their situation, resolved to try whether the enemy were equally inactive. Accordingly, Captain Urquhart, with Lieutenants Tyler, Macpherson, and Ouseley of the 45th, walked down the steep slope towards the advanced parts occupied by the enemy, and arrived at the spot whence the artillery had been withdrawn only a short time previously. Here they found some straw, which offered so strong a temptation to obtain a few minutes' repose that each ensconced himself beneath a heap, and prepared to enjoy his good fortune. They were soon fast asleep; even the roll of the drums was unheeded, and the first sound that broke their rest was the clash of bayonets. This ominous sound effectually roused them, and they scampered back to their regiment with admirable expedition—a retrograde movement which was accelerated by a strong impression that they could hear the enemy coming up the hill. Upon reaching their line they found the regiment formed, and silently waiting the attack. To fall in without being observed by the Colonel (Meade) was out of the question; they had long been missed, and he had sent orderlies in all directions after them; and he now pounced upon them, as they approached, full of indignation at this infringement upon military discipline. He loudly called to them—'There you are! I'll report every one of you to the General; you shall all be tried for leaving your ranks while in front of the enemy!'

"Observing at this moment that they were attempting to fall in and avoid further castigation, he assailed them with renewed eloquence—

"'Stop, sirs! stop!—Your names! for every one of you shall be punished; it's desertion!' and a great deal more he would have added, but that the French were on the move; and each officer having given his name, without waiting for any further observations, occupied his post in the ready, formed ranks, much chagrined at the unfortunate event of their expedition and its probable result. But the fight soon began, and every other thought was absorbed in the heat of battle. After the enemy had been repulsed the firing ceased, and the allies were falling back upon Coimbra. Colonel Meade, who was a severe disciplinarian, and possessed a most inveterately good memory, resolved to fulfil his promise, and report the offending officers to General Picton. Seeing Lieutenant Macpherson, he called to him,. and in a tone of severity said:—

"'Well, sir! you remember last night, I suppose?' Macpherson bowed with no very enviable recollections.

"'Ah! it's a breach of discipline not to be forgotten,' con-

tinued the Colonel, with a stern and uncompromising look.
'Where is Urquhart?'

"'Killed,' replied the lieutenant.

"'Ah!' grunted out the disciplinarian, 'it's well for him.
But where is Ouseley, sir?'

"'Killed, sir,' again responded Macpherson.

"'Bah!' exclaimed the Colonel, in a still louder tone, as
if actually enraged at thus being deprived of the opportunity
to punish their breach of military discipline.    As a last
resource, however, he enquired:—

"'Where is Tyler?'

"'Mortally wounded, sir,' was the reply.

"This was too much for the old Colonel's patience, so, with
a look of anger, not at all allied either to regret or repentance,
he rode off leaving his only remaining victim in a state of
much uncertainty.    Two days after this *rencontre* Lieutenant
Macpherson, having received a message from his friend,
Tyler, who, with the rest of the wounded, had been carried
into Coimbra, requesting to see him, he applied to Colonel
Meade for leave to visit the town, stating, at the same time,
that his object in doing so was to attend, as he thought, the
dying moments of his friend.    The Colonel had not, however,
forgotten Macpherson's offence, and he took this opportunity
to punish him.

"'No, no,' said he, in a voice which seemed to forbid all
further solicitation, 'you shan't go; you haven't deserved
it, sir.  Go to your duty.'

"Macpherson shortly after this met General Picton, and to
him he stated the request which his chum Tyler had made,
and Colonel Meade's refusal to grant him leave.  Picton was
indignant.  'What! not let you go?' he exclaimed, in his
usual forcible and energetic manner; 'damn me! you shall
go, and tell Colonel Meade I say so; d'ye hear, sir?'

"The young lieutenant both heard and obeyed.  Thanking

the General, he set off, first, to deliver Picton's message to the infuriated colonel, who swore that 'all discipline in the army had ceased'; and, then, to Coimbra, where he found his friend Tyler not dead, nor dying, but wonderfully recovered from the severe wound he had received, and prepared with an excellent breakfast for Macpherson and some more of his companions, whom he had contrived to allure into a participation of the good cheer he had provided by the invitation to attend his dying moments."

Masséna, after his defeat, was within an ace of retiring over the Portuguese frontier, but information of the road round to the northward of the English position having reached him, he determined to turn it by this route. Under cover of skirmishers he started for this purpose on the 20th of September. Towards evening, however, Wellington discovered his purpose, and straightway gave orders for the army to retire upon the lines of Torres Vedras. They marched by the same route by which they had advanced upon Roleia and Vimiera in 1808; their rear was in occasional touch with the French advanced guard, but by the 8th of October the British were drawing into the lines, and by the 10th and 11th the French began to recognise the nature of the fatal trap into which they had been so skilfully led.

Masséna lay idly watching the British in their lines until the middle of November, when from the difficulty of collecting food, and the increasing sickness following privation, he was compelled to commence his retrograde movement, and retired to Santarem, twenty-five miles to the north, on the bank of the Tagus, whither he was followed by Hill and Crauford. His position, however, was so well taken up that it was impossible to attack him; and, being in a better position to draw supplies from the surrounding country, he disappointed the allies by making a quiet stand when they had hoped he might be flying before them.

The gallantry displayed by the 45th at Busaco had the effect of saving the lives of four men of the regiment who had been sentenced to death for highway robbery, committed on some Portuguese inhabitants at or near St. Euphemia. On the 30th of September the following order by the Commander-in-Chief occurs:—" No. 3. Although the Commander of the forces has long determined that he will not pardon men guilty of crimes of which the prisoners have been convicted, he is induced to pardon these men in consequence of the gallantry displayed by the 45th Regiment on the 27th inst. (battle of Busaco). He trusts this pardon will make a due impression upon the prisoners, and that by their future regular and good conduct they will endeavour to emulate their comrades, who have, by their bravery, saved them from a disgraceful end."

In March, 1811, Wellington received considerable reinforcements, and in spite of the fact that Soult, in January, had defeated the Spaniards at Gebra, and laid siege to Badajoz, Masséna's position became untenable, and by the 6th of March his army was in full retreat upon the valley of the Mondego.

Wellington's army followed rapidly after the retiring French; the first, fourth, and sixth divisions followed directly by Thomar, and the third and fifth divisions marched on the west side of the hills; the whole, together with the cavalry and Pack's brigade, concentrated at Pombal, where the French appeared to be about to give battle on the 10th of March, and prepared to attack; but Masséna, after a slight skirmish, in which a lieutenant of the 45th was killed, continued his retreat along the left bank of the Mondego, leaving Ney to cover it by opposition at Redinha on the 13th, at Casal Nova

on the 14th, and at Foz d'Aronce on the 15th. In each of
these actions Picton's division had the duty of turning the
enemy's left, and at Foz d'Aronce were so rapid in their
movements that the enemy left most of his baggage in
their hands.

This harassing of the French continued until the 29th,
when they made a stand at Guarda, on which Picton called
"the strongest and most defensible ground I ever recollect
to have seen in any country."

They were again driven off, however, by Picton with his
third division, now and ever to be known in history as " The
Fighting Third," in its usual style; and on the 1st of April
the allied army descended from the mountains to the Coa
with a fixed determination to drive the French over the
frontier.

On the 3rd the French were again driven back at Sabugal,
but the third division were not seriously engaged, the 45th
having but two rank and file wounded; and by the 5th the
whole of Portugal was clear of the invaders, with the exception of Almeida, the garrison of which could be scarcely
expected to hold out long, as it was reported that they had
but ten days' provisions left.

In order to cover Almeida, and partly, no doubt, to
provision his army, Wellington went into cantonments in
support of the forces investing the fortress. Masséna, meanwhile, after reinforcing and supplying Ciudad Rodrigo, fell
back for a time to Salamanca.

During his advance Wellington had detached Marshal
Beresford and his Portuguese to the southward with a view of
operating against Badajoz and Andalusia. By the 25th of
March he had driven the French out of Campo Mayor, and
practically out of Portuguese territory, and proceeded to
invest Olivenza and Badajoz. Wellington, calculating on a
pause in the operations in the north, quitted his army on

the 14th of April, and, riding hard for three days, met
Beresford at Elvas on the 17th. He reconnoitred Badajoz,
gave directions for its siege, and, retracing his steps on the
25th, rejoined his army in the north on the 28th.

On his arrival he found that Masséna had again advanced,
and had reached Ciudad Rodrigo, with the evident intention
of trying to raise the siege of Almeida. Wellington
accordingly drew up his army on the westward bank of the
river Duas Casas, midway between the Coa and Agueda, with
his left resting upon the dismantled Fort Concepcion, and
his right behind Fuentes d'Onoro. The fifth division was at
Fort Concepcion, and the sixth and light divisions formed the
centre, opposite Alameda. The rest of the army, consisting
of the first, third, and seventh divisions, were massed behind
Fuentes d'Onoro, five battalions being placed in the village
itself, with orders to hold it. Such was the position in which
Wellington, with 32,000 infantry, 1200 cavalry, and 42 guns,
awaited Masséna with 40,000 infantry, 5000 cavalry, and
36 guns.

Masséna advanced on the 3rd of May, sending two corps to
threaten the British left and centre, while he kept Loisson's
corps, Drouet's division, and all his cavalry for an attack on
the right. Loisson, without orders, threw himself upon the
village of Fuentes d'Onoro, and fought a bloody battle for
its possession against the five battalions, of which the 45th
was one, of the third division, which held it. The English
troops were driven out of the lower town by the force of
numbers, but they held the high ground round the church,
and ultimately retook the village, driving the French out.

On the 5th the enemy, who had been reinforced, renewed
the attack, and succeeded in turning the British right. The
battle raged particularly fiercely in and round the village of
Fuentes d'Onoro; twice the French gained the village and
twice were they driven out again—on the first occasion by

the 88th and 74th, and on the second by the 79th regiment. The battalions retired in square before the French cavalry, and were unbroken; but the 45th, under Major Greenwell, received the French cavalry in line with such a determined aspect that they dared not face them, but wheeled about and retired.

Lieut.-Colonel Campbell says : —" Lord Wellington, I have reason to believe, ordered the 45th regiment (then under the command of the present Major-General Greenwell), such was his opinion of their firmness, at the battle of Fuentes d'Onoro, to receive in line, and without forming square, the enemy's cavalry then advancing in force towards them, if they should venture to charge. The experiment was not, however, made, for the French, I conclude, observing such a steady, determined front presented to them, thought it wiser to retire."

Though so far successful on one flank the enemy failed to make an impression on Fuentes d'Onoro, where they were resisted by the 45th, 74th, and 88th, supported by the 71st and 79th, and, not seeing his way to push his advantage, and ammunition failing, what would probably have been a rout with any but British troops, finished as a drawn battle. The British loss was 1200 killed and wounded, and 300 prisoners; that of the French probably much greater.

The 45th regiment had two rank and file missing on the 3rd, and 2 rank and file killed and one missing on the 5th; it was not, therefore, so heavily engaged as usual.

Finding the relief of Almeida hopeless, Masséna managed to pass a message to the Governor to blow up the works and retreat upon Ciudad Rodrigo; which, partly by his own skill and partly through the neglect of the British, he succeeded in doing on the 10th. On the same day Masséna passed the Agueda, and went into cantonments about Ciudad Rodrigo, where he was shortly superseded in the command by Marshal

Marmont, under whose direction the French army soon after withdrew to Salamanca.

Wellington, now anxious to clear both routes out of Portugal, divided his army into two parts, sending the third and seventh divisions and the 2nd German Hussars by forced marches southward to Badajoz.

On the 16th of May Beresford defeated Soult at Albuera, and on the 27th 5000 men were investing San Christoval and 10,000 surrounding Badajoz, while an Anglo-Spanish covering army was extended from Merida to Albuera to watch Soult.

On the 6th and 7th of June assaults were delivered on Fort Christoval, which failed; and news of the advance of Marmont and Soult to raise the siege forbade its continuance, and it was converted into a mere blockade, which, on the 18th of June, was altogether raised, and Wellington fell back upon Elvas and Campo Mayor, where he was joined by the remainder of the army from the north. The position chosen by Wellington was one which concealed his dispositions from the enemy, while every movement of the French was visible to him.

The French crossed the Guadiana on the 23rd of June, but though there was a slight skirmish no decisive engagement took place, and a diversion against Seville by the Spanish General Blake compelled Soult to fall back, and thus break up his combination with Marmont, who, finding he could do nothing by himself, fell back to the valley of the Tagus, on Truxillo and Zafra, guarding the bridge at Almara and still covering Badajoz.

Wellington's advance into Spain was still rendered impossible by his inability to take either of the great fortresses of Badajoz or Ciudad Rodrigo, which barred his way. Accordingly, leaving Hill to watch Badajoz and Soult's

F

army, he moved northward again, covering the advance of a siege train of 68 heavy guns, and reached his old position near the Coa about the 8th of August.

The allied position was moved close up to Ciudad Rodrigo, and the "Fighting Third," as usual, was in the foremost place. It occupied the heights of El Bodon, within three miles of the fortress, and overlooking the plain which surrounded it.

On the 25th of September the French Imperial Guards attacked the left wing of the allies, and drove in the outposts, but were afterwards checked and driven back. While this was going on Montbrun, at the head of some 16,000 troops with 12 guns, crossed the Agueda and marched straight up the hill to El Bodon; and Wellington, seeing that the position of the third division was untenable, sent them orders to retire on Fuente Guinaldo, where he had formed an intrenched camp.

Possibly there is no other occasion on which greater steadiness was shown by the "Fighting Third" than in this retirement across six miles of absolutely flat country without the slightest protection from any incident of ground, without supporting artillery, and practically without cavalry. Throughout the march the enemy's cavalry never quitted them, and six guns taking them in flank subjected them to a fearful fire of cannister and round-shot. Sometimes in squares, at other times in columns, they manfully resisted every effort of the French cavalry to charge home, and, with comparatively trifling losses, victoriously accomplished their retirement to Fuente Guinaldo.

Lieut.-Colonel Campbell, in writing of this march, again refers to the gallantry and steadiness of the 5th and 45th regiments in receiving charges of the enemy's cavalry, the 5th on one occasion even going so far as to charge the enemy's cavalry themselves, driving them down the hill with considerable loss.

The next day there was some slight fighting at Aldea Ponte, and on the 28th Wellington fell back into a position facing the enemy with his back to the Coa.

Marmont, however, was hampered by the difficulty of bringing up supplies, and once more turned his back upon Portugal. He re-garrisoned Ciudad Rodrigo in passing, and fell back upon Salamanca, both armies going into winter quarters.

Thus ended the campaign of 1811, in which the 45th, brigaded with the 74th and 88th, had their full share of fighting. They had well maintained the credit of the regiment, and had played their part in accordance with the best traditions of the British army. The *esprit de corps* which existed in the brigade was probably no small factor in the enthusiasm and steadiness with which it always fought. Lieut.-Colonel Campbell makes special allusion to it, and says :—" How often have I witnessed this feeling prevail to a great extent among the officers and soldiers of the 45th, 74th, 88th, and 5th battalion 60th which composed the brigade to which I was so long attached as brigade-major. I can never forget one instance of this kind in particular which occurred at Fuentes d'Onoro. After a long and dreadful struggle in that village between the British and a large body of the French Imperial Guard and other troops which supported them, the right brigade of the third division was at last brought up to take the place of the fatigued regiments so long engaged, under a heavy fire, without any decided advantage being gained by either party, the French holding the lower, and our troops the upper, part of the village. The 88th, supported by the 45th, was ordered to charge into it and drive out the enemy. They soon did so in the usual style

of the third division.   But I shall ever think with pleasure
of the extraordinary eagerness evinced by the 45th to advance
to the help of their old friends, the Rangers, who on that
occasion wanted none.   This feeling, however, between these
two corps in particular was always most strongly marked
throughout the whole war, and I have no doubt would be
revived if they ever met again in presence of an enemy."

# CHAPTER VI

WHILE the events described in the last chapter were proceeding the materials for the siege of Ciudad Rodrigo were being secretly collected at Almeida; while the town itself was watched and partially blockaded by Spanish guerillas.

The town stood on the right or north bank of the Agueda, and was connected with the south bank by a single bridge. The gate into the town was just opposite the end of the bridge, and close to it, the bridge being overlooked by a stone keep, or castle, somewhat higher than the other buildings. The ramparts stretched away east and west, being especially strong to the east, where the ground was high; while to the west was a partially-protected *faussebraye*, or outer rampart, between the counterscarp and the ramparts; further to the north in this direction was a hill called the Small Leson, overlooking the *faussebraye*, and still further on a greater height known as the Great Leson, crowned by a small, closed work, called *Francisco*, and garrisoned by 50 or 60 men. The ditch and ramparts ran right round the town, which was of somewhat triangular shape, till they reached the rocky heights close over the river. On the east side of the town there was a considerable suburb surrounded by entrenchments, while at the north end was the fortified convent of San Francisco and at the south end the convent of San Domingo, also fortified. The Salamanca gate gave entrance into the town on this face, and further south was the St. Jago gate. The town was garrisoned by some 1500 men, and contained immense stores of war material, including the whole of Marmont's battering train.

On the 15th of October the town lost its governor—Renaud —who was taken while endeavouring to rescue a convoy which had been attacked by the Spanish guerillas, and sent a prisoner into Wellington's camp.

Before the close of the year Wellington had thrown a bridge across the river at Marialon, about 6 miles below Ciudad Rodrigo, but the difficulties of transport were so great that it was the 8th of January before any forward movement was made. This delay made the rapid conclusion of the siege imperative, for it was impossible to say how soon Marmont might move up to the relief of the town. An assault, therefore, was from the first a foreseen necessity, and it was determined to deliver it from four points simultaneously.

Fort Francisco and the Great and Small Leson were to be besieged, and batteries erected on the high ground were to effect a breach in the extreme northern angle of the works, and a lesser breach on the north-eastern face. These were to be stormed simultaneously, while another body of troops assailed the St. Jago gate, and a smaller force—crossing the old bridge—was to endeavour to force an entrance below the castle.

Accordingly, early on the 8th of January, the light division under Crauford and Pack's Portuguese crossed the fords above the town, and, making a wide circuit round the east of the fortress, lay *perdu* for the rest of the day to the north of the Great Leson.

When night came, Colonel Colborne of the 52nd, with two companies from each battalion of the light division, stormed and carried Francisco fort, making 40 of the garrison prisoners, with a loss of 24 officers and men. The Portuguese then set to work to right and left of the captured fort, and

PLAN
OF THE FORTIFICATIONS
of
CIUDAD RODRIGO
Illustrative of the Sieges of
July 1810 & Jan. 1812

The field works refer to the Siege of 1812

SCALE

[To face p. 86.

by daylight on the 9th had driven a parallel some 600 yards diagonally across the Leson, giving approach and cover to begin the batteries when darkness should again come on.

The divisions of the army took it in daily turns to carry out the trenching work, and to protect the workers; and by the 13th, 28 guns were mounted in the batteries, the second parallel and its approaches were pushed on by flying sap, and the Germans of the first division carried the convent of Santa Cruz by surprise, and thus protected the left flank of the besiegers' works.

On the 14th the enemy made a successful sortie, doing some damage, but in spite of this, at half-past four that afternoon, 25 guns opened on the *faussebraye* and the ramparts, while two guns were brought to bear upon the convent of San Francisco. The enemy replied vigorously, but when night came the 40th regiment gallantly rushed the convent, and established themselves in the suburb. On the 19th, after four days' continuous bombardment, Major Sturgen, commanding the Royal Engineers, reported the breaches practicable, and the assault was accordingly ordered.

The direct assault upon the main breach was committed to the "Fighting Third" division and MacKinnon's brigade. The 45th, 74th, and 88th regiments were awarded the post of honour in front, the 45th leading, who were to follow immediately upon a party of 180 sappers, carrying bundles of hay to be thrown into the ditch, which was thirteen feet deep opposite the breach.

Whilst waiting for the hour fixed for the assault, an order arrived from Sir Thomas Picton to form a forlorn hope. The officers commanding companies were therefore called together and ordered to bring to the head of the column six men from each company for the purpose. They soon returned, declaring that every man present volunteered for the pre-eminence, and wished to know how they were to act, for the oldest soldiers claimed it as their right.    The moment for the assault had arrived, and there was no time to be lost, so Captain Martin, commanding the Grenadier company, who was there badly wounded already, put an end to all difficulties by requesting leave to lead as he stood with his company at the head of the regiment.    This was very reluctantly acquiesced in, but there was no time to make other arrangements, and the regiment advanced rapidly and in perfect order towards the breach.

As the 45th, stoutly supported by the 74th and 88th, rushed up the gentle slope of the glacis, the whole of the ramparts suddenly blazed with fire, and the deadly grape came tearing through the advancing ranks of the regiment, which, in no way daunted, soon gained the outer edge of the ditch; the sappers flung their bundles of hay over the counterscarp, and, impatient of an instant's delay, the stormers sprang down into the ditch, picking themselves up at the bottom as best they could.    On they went, scrambling over the crumbled masonry of the *faussebraye*, down into a second shallower ditch and up the wide and broken acclivity of debris which formed the ascent to the great breach.    Here they met the heads of the 5th, 77th, and 94th regiments, which had come along the ditch from the left, and up they all swarmed together, the way choked and tangled with crowds of dead and wounded, who, falling, hung upon the rugged blocks of stone, or rolled unheeded to the bottom of the ditch.

At this instant a loud and prolonged explosion took place: the enemy had sprung a mine at the foot of the breach, which, providentially, the regiment had scarcely reached, but which laid low the gallant leader of the brigade, Major-General MacKinnon, and many who were with him.

This did not, however, for an instant stop the progress of the gallant stormers, who, ascending the breach under a most destructive fire, soon gained the summit, only to find, to their horror, that the breach was completely cut off, and that there was no possibility of descending into the city.

The part of the parapet thrown down by the artillery fire was the salient angle of the northernmost bastion; it was quite evenly done, and when the men reached the top they stood on a space about one hundred feet wide, terminated on either hand by the broken ends of the parapet, and the terre-plein behind it stretching wide on either hand. So the storming party, instead of being able, as they naturally expected, to spread right and left along the terreplein until they found slopes leading down to the level of the town, found themselves at the top of a great wall, sixteen feet high, exposed to a murderous fire from the houses close to their front, and from a high bank nearly in their rear.

A short but fearful pause ensued, but not a thought of retreat entered the mind of a single man among those gallant stormers, when suddenly the brigade-major, Major Wylde, appeared, greatly animated, from the right waving the troops on in that direction; with loud cheers the men followed him. In their hurry to fall back the French had not had time to remove a few planks laid across the cut to our right, and over these the 45th, followed by their supporting regiments, rapidly swarmed. More planks were found on the other side, and in a short time the leading brigade of the "Fighting Third," by unflinching and persevering gallantry, and under a most appalling fire, had fought their way into Ciudad Rodrigo. Here they joined hands with the light division, who had carred the smaller breach with comparative ease, and were taking the defenders in flank and rear. The French were soon driven from the ramparts, and the fortress was ours.

Thus was Ciudad Rodrigo won after a siege lasting only twelve days, at a cost to the allies of 90 officers and 1200 men,

of whom 60 officers and 650 men fell at the assault, including
Major-General MacKinnon and Major-General Robert Crauford, the hero of Buenos Ayres, and the renowned leader of
the light brigade, who was shot gallantly leading his brigade
to the assault on the little breach.

The 45th, from its forward position, naturally lost heavily ;
3 officers, 1 sergeant, 1 drummer, and 12 rank and file were
killed ; 4 officers, 1 sergeant, 1 drummer, and 25 rank and
file were wounded. Among the officers killed was Captain
Hardiman, of whom it was said that "three generals and
seventy other officers had fallen ; yet the soldiers, fresh from
the strife, only talked of Hardiman." He was known all
through the army, and carried joy and light spirits wherever
he went. A friend of his in the 88th writes of him and of
Lieutenant Pearse, who fell by his side in the breach, as
follows : —

"At length I reached the grand breach ; it was covered
with many officers and soldiers. Of the former, amongst
others, was my old friend Hardiman of the 45th, and William
Pearse of the same regiment. The once cheerful, gay Bob
Hardiman lay on his back, half his head was carried away
by one of those discharges of grape from the flank guns at
the breach which were so destructive to us in our advance.
His face was perfect, and even in death presented its wonted
cheerfulness. Poor fellow ! he died without pain, and
regretted by all who knew him. His gaiety of spirit never
for an instant forsook him. Up to the moment of the
assault he was the same pleasant Bob Hardiman who
delighted every one by his anecdotes, and none more than my
old corps, although many of his jokes were at our expense.

" When we were within a short distance of the breach, as
we met he stopped for an instant to shake hands.

" ' What's that you have on your shoulder ? ' said he, as he
spied a canteen of rum which I carried.

"'A little rum, Bob,' said I.

"'Well,' he replied, 'I'll change my breath; and take my word for it, in less than five minutes some of the subs. will be scratching a captain's ——, for there will be wigs on the green.'

"He took a mouthful of rum, and taking me by the hand, squeezed it affectionately, and in ten minutes afterwards he was a corpse.

"The appearance of Pearse was quite different from his companion. Ten or a dozen grape shot pierced his heart, and he lay, or rather sat, beside his friend like one asleep."

The other officer killed was Lieutenant Bell, and the officers wounded were Captains Milne and Marten and Lieutenants Humphrey and Phillips.

Picton, in his divisional order dated 20th January, says:—

"By the gallant manner in which the breach was last night carried by storm the third division has added much credit to its military reputation, and has rendered itself the most conspicuous corps in the British army."

He specially mentioned the 45th, and named Captain Milne as peculiarly deserving of his thanks. Picton gave the Grenadier company of the 45th a hundred guineas for their share of the night's work, with the expressed hope that the men would do him "the honour to drink to the future success of the third division."

Lord Wellington in his despatch said:—

"The conduct of all parts of the third division in the operations which they performed with so much gallantry and exactness on the enemy on the evening of the 19th, in the dark, afforded the strongest proof of the abilities of Lieut.-General Picton and Major-General MacKinnon, by whom they were directed and led. . . . It is but justice to the third division to report that the men who performed the sap belonged to the 45th, 74th, and 88th regiments, under the command of Captain Macleod, R.E.; Captain Thompson, 74th regiment; Lieutenant Beresford, 88th regiment; and Lieutenant Metcalfe, 45th regiment; and they distinguished themselves not less in the storming of the place than they had in the performance of their laborious duty during the siege."

After the fall of Ciudad Rodrigo the troops went into cantonments for a time on the Coa; while Marmont, forced by want of provisions, scattered his army in Leon and Old Castile.  By the end of February, however, the army was once more on the move, this time to the southward, with Picton still at the head of the "Fighting Third," and the 45th, 74th, and 88th still brigaded together under Sir A. Kempt, who had succeeded the late Major-General MacKinnon.

By the 15th of March a pontoon bridge had been thrown across the Guadiana, about four miles from Elvas, and on the following day Marshal Beresford crossed the river, drove in the enemy's posts, and invested Badajoz with the third, fourth, and light divisions and a brigade of Hamilton's Portuguese, in all about 15,000 men.

Badajoz, situated on the east bank of a bend of the Guadiana, had been strongly fortified; the high ground to the eastward, which somewhat commanded the town, was crowned by a work called "the Picurina," while the bridge crossing the river was strongly protected by a *tête-du-pont*, and connected by a covered way with the strong bastioned fort of Christoval, opposite the Picurina; while between the Picurina height and the town ran a small stream of the Rivillas, which had been dammed, and formed a considerable inundation.  To the south-west of the town, on a height, rose the strong work of Pardaleras, which was connected with the body of the place by a covered way; while the castle, situated on high ground, was some little way to the north of

SIEGE
of
BADÁJOS
BY THE ALLIES UNDER WELLINGTON
From 17th March to 6th April 1812

SCALE

[To face p. 93.

the bridge; while near to the river, to the extreme south, was the formidable bastion of San Vincente.

As in the case of Ciudad Rodrigo, Wellington was compelled to push on the siege with vigour and secrecy, owing to the presence in the neighbourhood of a large force under Soult. It was, in fact, a race between the power of attack of the smaller British force against the capability for concentration and movement of the larger French army.

Badajoz was, however, much stronger than Ciudad Rodrigo; its garrison numbered 5000 men, and was commanded by General Phillipon, a leader almost unequalled for his energy, steadfastness, and fertility of resource.

On the night of the 17th of March the allies broke ground against Picurina, on the side furthest from the town, and by daylight had completed a parallel six hundred yards long, three feet deep, and three feet six inches wide, with communication some four thousand feet in length leading up to it. The parallel itself was not more than one hundred and sixty yards from Picurina. The next night it was prolonged to the right and left, and two batteries traced out. Wet and stormy weather harassed the workmen and flooded the trenches, but in spite of this the parallel was extended across the Seville road, towards the river, by the 21st, and three counter-batteries were commenced between Picurina and the river in order to open on San Roque, the *tête-du-pont* which covered the bridge and dam across the Rivillas, as well as the castle and the ground to the left of it. On the 23rd the floods in the trenches suspended all work, but on the 24th the fifth division invested the place on the right bank of the river, which had up to this been unoccupied; and the weather having cleared, the batteries were armed with ten 24-pounders, eleven 18-pounders, and seven 5¼-inch howitzers, all of which opened fire on the following day.

They were replied to vigorously, and one of the howitzers

was dismounted, while several artillery and engineer officers were killed. In spite of this the San Roque was silenced, and the Picurina garrison so galled by the fire of our marksmen that no one dared to look over the parapet; and the outward appearance of the fort showing no indications of great strength, General Kempt was ordered to assault.

The "Fighting Third" were, as usual, to the fore, and five hundred men of that division were assembled for the attack on the evening of the 25th. At nine o'clock, the evening being fine, flanking columns, each composed of two hundred men, moved out to the right and left, while one hundred men remained in reserve in the trenches. The two flanking columns, headed by the engineers with ladders and axes, advanced simultaneously against the palisades, but were held in check until Kempt sent forward the reserve under Captain Powis of the 83rd, when, in spite of a fearful fire, the stormers scrambled over the palisades and up to the ramparts. Here a desperate hand-to-hand fight raged; nearly all the officers fell, and half the garrison were killed or wounded before the commandant, Gaspar Thiery, surrendered with the remainder, numbering eighty-six. Thus the first phase of the attack was completed, and the Picurina won; three battalions were at once advanced to secure the work, and the second parallel was begun.

The breaching batteries opened on the 30th of March, and by the 6th of April three breaches had been made, all of which were considered practicable. The main breach was on the right flank of La Trinidad bastion, another in the curtain of La Trinidad, and the third in the left flank of the Santa Maria bastion.

The four divisions were employed in the assault. The light division was told off to attack the main breach; the fifth division was to make a false attack on Pardaleras, and a real attack on San Vincente; the fourth division was to attack the two smaller breaches in the bastions of La Trinidad and Santa Maria, and the "Fighting Third" was ordered to escalade the castle.

The night of the 6th April was cloudy but fine, and the assault was ordered for ten o'clock. Half an hour before that time Picton was in the rear of his division, resting himself, nursing a hurt he had got in camp, so as to be ready for the tremendous exertions that were anticipated. Before the appointed hour a carcass fired from the town disclosed the position of the third division, and drew fire upon it. It was useless to wait, and the "Fighting Third," led by Kempt in Picton's absence, rushed forward. The troops, headed by their ladder men carrying the largest ladders procurable, crossed the bridge over the Rivillas in single file under a terrible fire, and advanced up the rugged and broken ground which led to the foot of the castle walls. Here Kempt fell, severely wounded, and as they were carrying him to the rear they met Picton and his staff hurrying up.

In the advance, Kempt's brigade formed the right of the column, and the 5th, 77th, 83rd, and 94th regiments, under Colonel Campbell of the 94th, the left; while the light companies of all these regiments, with three companies of the 60th Rifles, formed the advanced guard, in rear of which the 45th led the whole column.

The light-balls of the enemy completely exposed the position of the third division, and the loss as the men swarmed up to the walls was terrible. The ladders, which proved too short, were thrown down as soon as they were raised, while stones, logs of wood, and all sorts of missiles made terrible havoc among the men. At length three ladders were fairly well placed, and up the first of them climbed Lieutenant Macpherson of the 45th, closely followed by Sir Edward Pakenham. He arrived at the top before he discovered that the ladder was too short by some three feet; so, pushing the head of it from the wall he called upon his comrades to hoist him, ladder and all, upwards. Thrown up thus above the rampart he was shot by a French soldier before he had time to collect himself, two of his ribs were broken, and he was unable to move either way. As Pakenham was endeavouring to pass him the ladder began to give way under them. "God bless you, my dear fellow!" he cried, "we shall meet again!" meaning they would only meet in the next world. The next minute the ladder gave way, and Macpherson fell insensible in the ditch.

Meanwhile, Ridge of the 5th Fusiliers, managed to erect a ladder where the wall was lower and an embrasure offered a chance of entry, and a second ladder being placed by Cranch of the 88th, the two swarmed up over the parapet, followed by their men. The castle was won, and the troops crowding in, in increasing numbers, speedily drove the French in a desperate hand-to-hand conflict through the double gates into the town.

Picton had been shot in the groin, at the foot of the ramparts, and lay insensible for twenty-five minutes, but afterwards rose up and cheered his men on to the attack.

Macpherson, too, presently recovered himself, and, wounded and bleeding as he was, reclimbed one of the ladders into the castle, and tried to make his way to the keep where the French flag was flying. He seized the sentry, and, making his way to the flagstaff, hauled down the flag and hoisted his own jacket in its place, which, bravely fluttering in the breeze at daylight, testified to the gallant part the 45th had taken in the assault.

Meanwhile, the attack on the main breaches was going on with desperation, but with little success. For upwards of two hours a fearful fire of musketry and grape, together with hand grenades, bags filled with powder, and every conceivable form of destructive missile, had been poured on the heads of the attackers; and soon after midnight, when over 2000 men had fallen, Wellington ordered the remainder to retire and re-form for a second assault.

To the south and west of the town the 5th division had been more successful, having carried San Vincente by escalade, and moved across the town towards the great breach:

As soon as he was sure of the castle, Picton sent his aide-de-camp, Tyler, to Wellington to report, who sent him back to Picton with orders to hold the castle at all hazards. Feeling himself secure, Picton then sent parties to the left, along the ramparts, to fall on the rear of those defending the great breach, and to communicate with the right attack on the bastion of La Trinidad. But the enemy, upon retiring from the castle, had closed and strongly barricaded the gates communicating with the ramparts, and to force these barriers was a work of considerable time and difficulty.

Meanwhile, the French made a desperate attempt to retake the castle, in repelling which the gallant Brigadier of the 5th was killed.

The capture of the castle, however, coupled with the advance of the 5th division through the streets, soon convinced the French that it was useless to continue the struggle at the main breach; they gallantly rallied for some time and faced the troops taking them in rear, then broke and fled by the bridge and covered way into the fort of Christoval. When the stormers mounted the great breach for the second time they accordingly found the ramparts abandoned; and the troops, pouring into the unfortunate town from the three entries they had gained, were quickly masters of the place. Then followed a scene of riot and debauch unequalled in the annals of the British army; the chains of discipline were thrown off, and the

G

whole force gave themselves up to pillage, intoxication, and wanton destruction of life and property.

The first man who sprang down from the ramparts into the castle was Corporal Kelly of the 45th, who killed a French colonel as he did so; while the presence of Lieutenant Macpherson's jacket on the flagstaff adds evidence in support of the long-standing claim of the regiment, that it was the first to enter Badajoz.

The gallant Phillipon, with the remnant of his brave companies, surrendered Fort San Christoval on the following morning, and Badajoz was wholly won, but at the fearful cost of 5000 men, of whom 3500 fell in the assault.

Five generals were wounded—Kempt, Harvey, Bowes, Colville, and Picton; while of the "Fighting Third" no less than 600 officers and men were placed *hors-de-combat*.

It is said that when the losses were made known to Lord Wellington "the firmness of his nature gave way for a moment, and the pride of conquest yielded to a passionate burst of grief for the loss of his gallant soldiers."

The loss suffered by the 45th was as follows:—From the 18th to the 22nd of March, 2 killed and 10 wounded; from the 22nd to the 26th, Lieutenant Atkins and 7 men killed, Captain Lightfoot, Lieutenants Metcalfe, Marsh, and Andrews, 2 sergeants and 35 rank and file wounded, and 3 missing; from the 30th March to the 2nd April, 1 sergeant and 2 men wounded; and in the assault on the night of the 6th of April, Captains Herrick and Powell, Lieutenants White and Jones, Ensigns Macdonald and Collins, 1 sergeant, and 18 rank and file killed; Captains O'Flaherty and Reynolds, Lieutenants Macpherson, Dale, Monro, and Hill, Ensigns Stewart and Grant, 8 sergeants, 1 drummer, and 55 rank and file wounded; making a total of 171 casualties.

Macpherson on the following day took the flag he had captured to Picton, who, desirous of helping him forward, told him to take it direct to Wellington, which, after some hesitation, he did. Wellington thanked him, and asked him to dinner, an invitation which the state of his wound prevented his accepting. Two years later we find him still a *subaltern!*

# CHAPTER VII

AFTER the capture of Badajoz the news that Marmont had pushed past Ciudad Rodrigo into Portuguese territory at the head of 35,000 men induced Wellington, after leaving 10,000 men, under Sir Thomas Graham, to hold Badajoz, to march northward to his old cantonments between the Agueda and the Coa. Marmont, upon being advised of the British advance, retired upon Salamanca, and both armies settled down to prepare for the next move in the game.

By the beginning of June all Wellington's preparations for an advance upon Salamanca were completed, and on the 13th of that month he put his forces in motion and crossed the Agueda, arriving in front of Salamanca on the 16th. The passages of the Tormes were secured both above and below the town, and on the following day the third division crossed the river, and encamped on the north side of the place.

The only defences of Salamanca were certain forts inside the limits of the town; they were understood to be of no great strength, and the battering trains, which followed the army, were soon brought to bear on them. But though much damaged by bombardment, the first attempt to carry them by escalade failed; causing a delay which enabled Marmont to concentrate some 36,000 men and return and face Wellington's army, which was concentrated on the heights of San Christoval, some five miles beyond the town.

The third division was still composed of practically the

same battalions which had already so gallantly earned for it the title of the "Fighting Third"; its exact composition before Salamanca was as follows :—

> Right brigade, under Major-General Sir Thomas Brisbane: 45th, 74th, 88th, and 5th battalion of the 60th regiment.
>
> Left brigade, under Major-General the Hon. Sir Chas. Colville: 5th, 83rd, 87th, and 94th regiments.
>
> Centre brigade, under Major-General Sir Manley Power: 9th and 21st Portuguese regiments, and a battalion of Caçadores.

Artillery, a brigade of 9-pounders under Major Douglas.

On the evening of the 20th the two armies were facing each other in battle array, but nothing beyond small skirmishes occurred on the 21st and 22nd. On the 24th Marmont crossed the Tormes at Huerta, beyond the British right, and drove back the German cavalry. Wellington at once sent Graham, with 12,000 men, across a bridge upon his right, and Marmont was compelled to recross the river and fall back upon his old position.

On the 30th, after destroying the works at Salamanca, Wellington pursued Marmont towards the Douro, and encounters were frequent between the British advanced and the French rear guards. By the 2nd of July both armies were facing each other on opposite banks of the Douro, neither being strong enough to attack.

The third division at this time suffered an irreparable loss; Sir Thomas Picton was so completely struck down with fever that he had to be invalided home for the recovery of his health, and his place was taken by Major-General Pakenham.

The third division was at this time quartered near the deep ford of Pollos, and was opposed by the French seventh division; between them a close intercourse was established, and when the order to move was issued they parted with mutual expressions of esteem, and with hopes of a fair field in which it might frankly be seen who was the master.

Marmont, having been reinforced, at length considered himself strong enough to attack. Accordingly, he crossed the river at Pollos and Tordesillas, and gradually drove back the allied forces on Salamanca, so that by the 21st of July the two armies were once more facing each other in their old positions, Wellington's forces again occupying the heights of San Christoval.

The first contest on the 22nd was for the heights of Los Arapiles, which the French moved out in force to occupy; while the third division crossed the Tormes, and took up a position in the rear and on the right of the army. Later on the position of the main line was altered, and, instead of facing to the north-east, it faced south, with its left resting on the Arapile height, which still remained in the hands of the British, and its right joining the third division, now no longer in the rear. The nature of the ground was such that the enemy could not observe this change of front, nor be aware that any movement to the left would have to be along the British front, instead of round the flank.

About two o'clock, Wellington, who had retired to take a short rest, was roused by his aide-de-camp, who reported that the enemy were in motion. "Very well," said Wellington, "observe what they are doing." A minute or two elapsed, when the aide-de-camp said, "I think they are extending to the left." "The devil, they are," said his lordship, springing to his feet; "give me the glass quickly."

Gazing through the glass for a short time, and scrutinizing their movements, he at last lowered it, exclaiming, "At last I have them!" and turning to the Spanish general, who stood beside him, he said, "Mon cher Alava, Marmont est perdu!"

BATTLE
OF
SALAMANCA
22 July 1812.

At the same time he ordered Pakenham to move off with the third division and take the heights in front. "I will, my lord, by God!" replied the gallant soldier, "if you will give me a grasp of that conquering right hand"; and, parting with a hearty hand-shake, he proceeded to lead the advance of his division, which was destined to play so decisive a part in the great victory which followed.

When the order to advance arrived the men of the 45th were employed cooking their dinners; the camp kettles were at once overturned, packed on mules, and sent to the rear. The division was soon under arms, and moved rapidly off in open column, right in front, the 45th leading, the right brigade being under the temporary command of Colonel Wallace of the 88th.

The division moved at first into a hollow, the left brigade, headed by the 5th Fusiliers, marching parallel to the right brigade, so as to be ready to form a second line. The Portuguese brigade followed the right, and the whole of the left flank of the columns was covered by a cloud of skirmishers, composed of the light companies and the fifth battalion of the 60th Rifles.

At length, having fairly outflanked the French left, the whole formed line, and, led by Sir Edward Pakenham, hat in hand, advanced up the hill against the French position, from which a heavy and destructive fire was at once poured into the advancing British. But nothing could stop the "Fighting Third," and the French were charged and overthrown. But, just at this critical moment, the French cavalry, in turn, charged down upon the right flank of the 45th, but they found the "Old Stubborns" ready for them, who, ably assisted by the 5th Fusiliers, poured such a well-directed fire into their ranks that all apprehensions from that quarter were quickly removed; and the enemy's infantry were pursued with great impetuosity by Colonel Wallace, at the head of the Connaught Rangers.

The division, being re-formed, soon continued its advance under cover of a destructive fire from Major Douglas' guns. Another charge was intended, but the French would not stand, and retired in tolerable order, severely galled by our sharp-shooters, who were close at their heels. They then took up another position, in which they were reinforced by a large body of troops and a considerable number of guns.

Again Sir Edward Pakenham, bare-headed, placed himself at the head of the division, which, with an eagerness that could scarcely be restrained, advanced against the formidable position the French had taken up. Once more their victory was complete, and the foe, who would not wait to stand the charge of the " Fighting Third," were completely overthrown. Their broken columns were charged by our dragoons, who rode through and through them, cutting down numbers and taking thousands of prisoners, while the remainder, in total *deroute*, were running as fast as their legs could carry them.

Thus was the enemy's left completely routed by the gallant third division, aided and supported by our dragoons; a rout which carried alarm and confusion amongst the centre, and even as far as the right of their line.

A story is told of a private of the 45th who was after a French soldier with his bayonet; as a last resource, the Frenchman threw away his musket, and endeavoured to escape by climbing a tree, with the only result, however, that he received the bayonet in that portion of his person which should never be presented to either friend or foe.

Somewhat early in the action Marmont was severely wounded by a shell, and the command of the French army devolved upon General Bonnet, who held his ground on the right and right centre with considerable obstinacy.

The fourth and fifth divisions, supported by the sixth, seventh, and light divisions, had attacked the enemy in front, and the fight raged hotly round the Arapile height, which the enemy still held. Pack's Portuguese brigade assailed this height in vain, and the French right, where Foy was posted, was far enough in advance to shell our reserves; while Clausel was able to strengthen the centre, and offer a *point d'appui* for the broken left to rally upon somewhat.

Pack's failure, and this reinforcement of the centre, enabled Clausel for a time to take the offensive; he charged down upon the fourth division, which gave way—Leith, Cole, and Beresford all fell desperately wounded; but, Wellington, hurrying forward his reserves, soon restored the allied ascendancy at this point. Clausel then abandoned the Arapile height, and, covered by Foy's division, retreated on Alba de Tormes, pursued, in the gathering darkness, by the third division. The troops, however, were too worn out by marching and fighting to continue the pursuit and follow up the retreating French. But the victory was complete, nevertheless; and the 45th should never forget that it was the leading corps of the division that struck the keynote on that triumphant field.

The total allied loss amounted to 7264, out of some 46,000 engaged, of which the British casualties were no less than 694 killed, 4270 wounded, and 256 missing.

The 45th, which was commanded through the engagement by Lieut.-Colonel Ridewood, lost 5 rank and file killed; Major Greenwell, Lieut.-Colonel Forbes, Captain Lightfoot, Lieutenant Coghlan, and Ensign Rey, 1 sergeant, and 44 rank and file wounded. Major Greenwell was shot through both arms and the body by one bullet; he fell, as if killed, from his horse, but recovered, and lived to rejoin the regiment two years later as Lieut.-Colonel; he also received the C.B. for his conduct on the occasion.

It is related of the regiment that, in the first advance, the men could with difficulty be restrained by their officers, who kept pressing them back, until the word came from Pakenham, "Let 'em loose!" when they rushed like madmen on the bayonets of the French seventh division.

During the charge by the enemy's cavalry, previously related, the Grenadier company faced the charge and delivered 'a volley at close quarters which caused the French to wheel about and retire. Pakenham, who was riding past, exclaimed, "Well done, 45th!"

At one period during the advance, just as the enemy began to give way, a French officer seized a musket and attempted to shoot General Pakenham. The piece missed fire, whereupon the officer threw it down and picked up another; Corporal Cavanagh, of the 45th, at once sprang forward from the ranks and shot him dead, but was himself killed almost at the same moment.

The French, after the battle, retreated upon Arevalo, and King Joseph, who had advanced from Madrid at the head of 14,000 men, after nearly effecting a junction with Clausel, fell back again on Madrid, and passing through the city, retired to the southward. Wellington, instead of pursuing Clausel, turned aside after King Joseph, and occupied Madrid amid universal signs of enthusiasm and delight on the part of the inhabitants, on the 15th of August. The 45th, on this occasion, was awarded the place of honour, and marched at the head of the army in its triumphal entry into the city, the band playing "The British Grenadiers."

Clausel, having somewhat recovered from his defeat by the end of August, again advanced and re-occupied Valladolid, threatening our communications from Ciudad Rodrigo eastwards. Wellington accordingly marched out of Madrid again on the 7th of September, leaving the third and fourth divisions, as those which most required rest, to garrison the Spanish capital.

The advance of Soult and King Joseph from the south-east, in the autumn compelled Wellington to order the retreat of Hill's forces and the garrison of Madrid upon Salamanca. Accordingly, on the 1st of November, the third division was moved out to the passes of the Guadarrama mountains, north 'of Madrid. The British forces were hard pressed by Soult during the retreat, but by the 6th of the month the allied army was once again in its old position at San Christoval. On the 11th Soult was facing Wellington with a vastly superior force, and on the 14th the allied forces fell back upon Ciudad Rodrigo, covered by the cavalry and the light division. The town was reached on the 19th, and was occupied quietly, together with the neighbouring villages, and the campaign of 1812 closed.

# CHAPTER VIII

THE campaign of Salamanca was generally looked upon as a failure, yet the great victory which had signalised its opening was only shorn of its advantages by the retreat from Burgos. As a matter of fact, it had freed Andalusia and Estremadura, as well as the northern and western provinces of Spain, from the presence of the invaders; it had, moreover, stirred up an insurrection in northern and eastern Spain which threatened the French line of communications, and, above all, had put fresh heart into the cause of the allies.

By April, 1813, a great plan of campaign was matured by Wellington, having as its object the convergence of all the forces quartered in the western and southern parts of Spain upon the north-east corner of the country, the possession of which would command the main communications with France, and have the effect of drawing the French forces from all their positions which could not be held; while Biscay, Guipuzcoa, and Navarra were in possession of hostile forces.

In the beginning of May Wellington was prepared to take the field with some 200,000 men, a far larger force than had ever yet been assembled since the war began. About half his force, it is true, were Spanish, upon whom too much reliance could not be placed; at the same time, they were capable of rendering most essential service in keeping open communications, guarding convoys, blockading fortresses, and cutting off foraging parties of the enemy, thus leaving the Anglo-Portuguese force in undiminished strength to maintain the serious conflict in the front of the advance.

The earliest movements of the allies in this year were in the south-eastern parts of Spain, and when these began to develop, and to have the anticipated effect of blinding the enemy as to the true line of advance, Wellington began to send his cavalry across the Douro into Tras-os-Montes. They were followed by the pontoon train, artillery, and infantry, until the whole of his left wing, amounting to about 40,000 men, under Sir Thomas Graham, was in movement north of the Douro.

This first movement of the army, about the middle of May, sketched the strategy intended to prevail throughout the advance. The left wing, always operating in advance of, and on the left flank of, the right, was to be ever a turning force to every position which the enemy might take up. Thus it was calculated that, however anxious the French might be to offer battle to Wellington in their front, Graham's appearance on their right rear would force the abandonment of the design.

The " Fighting Third " was once more well to the fore with Graham's force (which comprised also the first, fifth, sixth, and seventh divisions), and again under their old leader, Picton, who had rejoined the army thoroughly restored in health.

On the ·22nd of May, Graham being well advanced, Wellington put his right wing in motion along the well-known roads towards Salamanca and the Tormes.

On the 29th the French reconnoitring across the Esla, came into contact with Graham's advanced troops; and, imagining he had the whole allied army in front of him, Rielle, who was in command, recrossed the river, and broke down the bridge behind him, leaving only some light cavalry to watch it.

Meanwhile, on the 26th, the heads of Wellington's and Hill's columns were approaching the Tormes. Villette, the French commander, remained too long in position, and suffered some loss in his retreat; the allies sweeping on towards Zamora and Toro. On the 28th Wellington crossed the Douro in a basket slung on a rope stretched from bank to bank, hundreds of feet above the river, and joined Graham at Carrajales, on the right bank of the Esla.

The river presented more difficulties than had been anticipated, and it was not until the 31st that a body of Hussars, with infantry holding their stirrups, effected a passage, and surprised the watchers at the broken bridge. Then the pontoon was thrown across, and, the army advancing, entered Zamora the next day, the French falling back upon Toro, where their cavalry made a futile attempt to stand, and were driven back with loss.

The whole army was soon to the north of the Douro, and on the 4th of June Wellington marched forward by several parallel roads, Graham's force being still on the left front. On the 12th the French rearguard blew up Burgos, and, continuing their retreat, eventually took up a strong position along the north bank of the Ebro, from Aro, on the left, to Frias, on the right, covering Vittoria and the road into France by Pampeluna, where they were reinforced by 14,000 men under Clausel.

Wellington followed up closely, and leaving the sixth division at Medina de Pomul to guard his stores and supplies, pushed on, with almost superhuman exertions, over the mountains in pursuit. Graham with the first, third, and fifth divisions, swung round towards Vittoria from the north-west, and on the 18th drove back Rielle, who was concentrated between Orduna and Espejo, on the Bilbao road; meanwhile, the light division had come up with Moncane's division, and driven it back in confusion, taking 300 prisoners and a considerable quantity of baggage. On the following day Graham pushed forward to Jocano, while Lord Wellington, with the main body of the army, advanced to Sabijana, on the left bank of the Bayas.

Wellington's plan for the forthcoming battle was to assemble his main army along the line of the Bayas river, facing the French line of battle, between which and the allied army intervened, first, the Zadorra river, and then the Morillas mountains on the right, with three lower hills in the centre and to the left, over which Rielle had been driven, and which were to form the principal line of advance for the allies. Hill was on the right of the allied line, nearest to the bridge over the Zadorra, at Puebla. The light and fourth divisions, with the cavalry, were in the centre; then came the seventh division, and on the extreme left the "Fighting Third," led by Picton, with the 45th on its right.

Vittoria was several miles in rear of the foremost French line, and there was no way of retreat open to them except through the city; while the best road was the royal causeway, leading to the north-east, which could easily be got at from the road leading to the south-east from Orduna. The region about this road had already fallen into Graham's hands, and he was ordered to advance towards Vittoria along that road, and possess himself of the bridges across the Upper Zadorra. It was obvious that, should Graham's turning movement succeed, the French would be compelled to retire and secure their rear and lines of retreat.

At daybreak on the 21st, the weather being wet and misty, the centre and left of the allies advanced across the rugged hills in their front. Hill's division was moved out to the right to force the Puebla defile and attack the French left, in which he succeeded admirably. By ten o'clock he had carried the village of Puebla, and after some further fighting he drove the French back into the basin of Vittoria and occupied the village of Sabijana, immediately in front of their left.

Wellington's main attack had to wait until the third and seventh divisions had developed their turning movement against the French right over very arduous and difficult ground. While waiting, he was informed by a peasant that the French had left the Tres Puentes bridge unguarded. He at once sent Kempt's brigade to cross the bridge and take the enemy, who were watching the Villodas bridge in rear. The brigade crossed the bridge at a run, and Kempt, having occupied the required position, and finding he was not seriously attacked, called up the 15th Hussars and held his ground.

Meanwhile, the third and seventh divisions, having pushed on round the enemy's right, were in full cry for the Mendoza bridge. The enemy now brought up a considerable force of artillery, accompanied by cavalry and light troops, to oppose their advance; but Kempt's force, whose position had placed them on the left flank of this force, held them securely, and enabled Graham's men to push on undeterred.

The following story, related of Picton, must have occurred just previous to the advance of the third division, with Graham's force on the left:—It appears that while the advance was proceeding on the right, his men, and especially the 45th, chafed a good deal at the delay, and he had some difficulty in restraining them. About noon, their inactivity still continuing, Picton became furious, and exclaimed, "Damn it! Lord Wellington must have forgotten us!" At length an aide-de-camp rode up to Picton and asked him if he had seen Lord Dalhousie, who commanded the seventh division. Picton, who expected that he was bringing orders for the third division to move, was much disappointed, and

BATTLE
OF
VITORIA
21st June 1813.

Cavalry   Infantry   Artillery (Allies)
                     Artillery (French)

SCALE

answered sharply, "No, sir! I have not seen his lordship! But have you any orders for me?"

"None!" replied the aide-de-camp.

"Then, pray, sir!" continued the irritated general, "what are the orders you do bring?"

"Why!" replied the officer, "that as soon as Lord Dalhousie shall commence an attack on that bridge the fourth and sixth are to support him."

Picton could not understand the idea of any other division fighting in his front, and said to the astonished aide-de-camp, with some passion, "You may tell Lord Wellington from me, sir, that the third division, under my command, shall, in less than ten minutes, attack that bridge and carry it. The fourth and sixth divisions may support if they choose." Then turning to his division, he quickly put them in motion towards the bridge, exclaiming, "Come on, ye rascals! Come on, ye fighting villains!"

Before the descent to the bridge, the third division had gained a height from which the whole field of battle was displayed. Below them the Zadorra flowed from left to right, and immediately beyond it rose the high ground occupied by the French right, whence came a heavy fire from the guns posted there.

Far away on the left the smoke of Graham's attack was visible, and a little on the left front the roofs and domes of Vittoria showed themselves; while the intervening space was filled with a confused mass of French troops, military trains, waggon loads of plunder, camp followers, and flying adherents of the French flag.

The right brigade of the division, under Brisbane, composed of the 45th, 74th, and 88th regiments, passed the Mendosa bridge without opposition, the 45th being the first across; Colville's brigade passed by a ford higher up, and was immediately hotly engaged.

H

The Connaught Rangers, after crossing the bridge, diverged to their left, ascended the hill, and fell on the French occupying it. The remainder of the brigade, followed by Kempt's brigade, moved in front of the French to near the centre, where, joining the cavalry and part of the light division, they turned to the left, and charged up the hills into the middle of the French line, which made no stand, but quickly retired under cover of their artillery and a cloud of skirmishers.

On the range of heights in front of Gomecha, where their reserves were posted, and in the village of Ariniz, on the main road, the enemy endeavoured to make a stand; but Picton's troops, plunging into the streets amidst a heavy fire, quickly captured three guns. The post was important, and, French reinforcements arriving, the battle raged furiously for a while, but the issue was never in doubt, and, finally, the British troops issued forth victorious on the further side.

From this onward the fight was mainly a confused and disjointed assemblage of separate combats; the French vainly endeavouring to stand, and the allies always successfully pressing onward. About six o'clock the enemy made a last desperate effort to hold a defensible height about a mile in front of Vittoria; here they were desperately assailed by the " Fighting Third," with our gallant regiment, as usual, in the thick of it. Nothing could stop them, and the French were driven pell-mell from their last position, and, leaving Vittoria on their left, fled headlong along the road to Pampeluna, losing all their guns, stores, treasure, and other necessaries.

Thus was gained one of the greatest and most decisive victories fought in the Peninsula, which cost the allies 740 in killed, and 4170 wounded. The 45th had 4 rank and file killed; and Lieut.-Colonel Ridewood, Lieutenants Russell and Little, Ensign Edmunds, 5 sergeants, and 61 rank and file wounded. Lieut.-Colonel Ridewood died the following day, and the following brigade order was issued by Major-General Brisbane:—

"The Major-General cannot help lamenting that the 45th and the service should have been deprived of the able assistance of so valuable an officer as Lieut.-Colonel Ridewood, in consequence of his wounds, received at the termination of so glorious an action."

Picton, writing of the battle, complained that the *Gazette* account was "a most incorrect relation of the circumstances of that memorable event; most uncandidly attributing to arrangement and manœuvre alone what was in a very considerable degree effected by blood and hard fighting. The arrangements and combinations preparatory to the action were certainly excellent; but the centre of the enemy's army did not immediately fall back upon Vittoria on the seeing the arrangements for its attack, as represented in the official despatch, but, in fact, disputed every inch of the ground, and was driven from several strong positions by the third division alone, and with a loss in killed and wounded of 89 officers, 71 sergeants, and 1475 rank and file; a number which exceeded one-third of the whole casualties of the army on that memorable day, and, being in the same proportion to our own effective numbers, which were under 5000. Upon the whole, the division has not had its proportion of credit; but its operations were in view of the whole army, and murder will out in the end."

The battle having been gained, the necessity of immediate pursuit at once presented itself to Wellington. Accordingly, the sixth division was at once called up from Medina Pomar to occupy Vittoria, Graham was despatched in pursuit of Foy, and Wellington, with the rest of the army, including the third division, pursued Joseph's army, now in a state of destitution and insubordination, through Sabatiena towards Pampeluna.

Learning, however, that Clausel was approaching Vittoria by roads south of the Pampeluna road, Wellington left Hill to invest Pampeluna, and marched by Tafalla, on the road to Tudela, with two brigades of light cavalry and the third, fourth, and seventh divisions, for the purpose of cutting off Clausel from the east of Spain. But Clausel was too wary to be caught, and, marching to the south-east along the bank of the Ebro, he covered 40 miles in sixty hours, and escaped to Saragossa, which he reached on the 1st of July, and subsequently secured a position at Jaca, 20 miles from the French frontier.

Meanwhile, this attempt to capture Clausel had stopped the pursuit of Joseph, who, finding he was not pressed, attempted to hold the valleys of Roncesvalles and Bastan, which were rich in supplies; but O'Donnell's Spaniards having joined Hill, enabled the English general, without abandoning the siege of Pampeluna, to move forward sufficient forces to clear the French out of these positions, which left the allies in possession of the whole frontier line from Roncesvalles to the mouth of the Bidassoa, while Pampeluna and St. Sebastian, the chief points left to France in this part of Spain, were closely besieged.

Soult was now appointed to the supreme command of the French army, which he quickly reorganised, and also set about strengthening the defences of Bayonne; and, on the 16th of July, arranged his army—which consisted of 114,000 men—for a general advance.

Wellington's army consisted of about 82,000 men, and his lines extended from the passes of Roncesvalles to the sea.

At daybreak on the 25th of July Soult advanced against the passes, and commenced the long series of encounters known as the battles of the Pyrenees. On the 25th and 26th

the allies were compelled to fall back, but on the evening
of the latter day Cole, being supported by Picton, held his
ground for the night on some heights about one-fourth of the
way from the passes to Pampeluna.

In the meantime, Drouet had been advancing on Hill's
position, far to the left of Cole, by the passes of Maya,
leading into the Bastan valley. Stewart, who commanded
there, was forced back down the valley, but Drouet did not
follow up his advantage as he might have done. This delay,
as well as Soult's hesitation in attacking Cole, enabled the
allies to improve their condition by further concentration
in stronger positions. Wellington, who had come across
from St. Sebastian, on the 27th found Soult extending his
right and left, preparatory to an attack on Cole, who occupied
a position on some heights, with Picton in support on his
extreme right. The cheers which greeted Wellington's
arrival among his troops probably reached the ears of Soult,
who guessed their cause, and delayed his attack until the
next day, by which time another 6000 men had been brought
into the allies' line of battle. The attack, however, was in
vain, and, after a battle—characterised by Wellington as
" bludgeon work "—in which the French lost two generals
and 1800 men and the allies 2000 men, the repulse was
complete, and the British held their ground, with their left
and communications secure.

On the 20th the French began to fall back, and Wellington directed Cole to make an attack on the strong position
of Samoren, held by the French right, under Foy; while
Picton, who had been inactive on the extreme right of the
allied line, was ordered, with additional cavalry and
artillery, to undertake the task of turning the French left.
After a bloody battle, in which the French lost more than
2000 men, with 3000 prisoners, and the allies 1500 men, the

enemy abandoned all their positions and retreated up the valleys towards France.

It was now practically all over with Soult. Clausel in vain tried to sustain positions at Echalar, but was beaten out of them in confusion by Wellington on the 2nd of August, and on the following day the armies rested, facing each other on the frontier, over which Soult was practically driven. There had been nine days of almost continuous fighting, during which the allies had lost 7300 men in killed, wounded, and missing, and the French probably at least 13,000.

During these movements the third division, under Picton, had followed Foy over the frontier, and now looked down on France from the summits of the pass of Roncesvalles, on the right.

Wellington now applied himself to the capture of San Sebastian. The renewed siege was begun on the 5th of August; on the 31st the town was carried by assault after terrible slaughter; but the garrison, retiring into the citadel, maintained themselves there until the 9th of September.

While the town was being stormed on the 31st of August Soult crossed the Bidassoa and attacked the covering army. Wellington, however, had good information of his intentions, and made such dispositions as repulsed the attack at all points, Soult retiring to his former defensive positions.

After this there was again a pause in the operations; but, having received reinforcements of three battalions, and the brigade of Guards which had been left behind at Oporto, Wellington was, by the 7th of October, again ready to proceed. On the morning of that day seven columns, covering a front of five miles, began crossing the Bidassoa. The French were utterly surprised all along their line, and fell back across the Nivelle, taking up a strong position along that river, which Soult strengthened by every means in his power, as well as the works covering St. Jean de Luz, at its mouth.

The sufferings entailed upon the allied army by the growing severity of the weather, and the difficulty of procuring supplies in a rugged and mountainous country, devoid of good roads, were perhaps more severely felt by the 45th regiment, and the rest of the third division on the extreme right, than any other portion of the army. The regiment was at this time encamped above the village of Zugarramundi, between the passes of Echalar and Maya. On these heights the division was exposed to incessant rain, alternating more and more frequently, as the season drew on, with snowstorms. Provisions were painfully short, and the piquet duties, from the near presence of the enemy, were extremely heavy and harassing. It was a common saying in the division that all future hardships and privations must be regarded as trifling in comparison with the miseries of Zugarramundi.

The news of the fall of Pampeluna on the 31st of October prepared Wellington for an immediate advance; but his intentions were frustrated by bad weather, and his army could not be set in motion before the 10th of November, on which date, at the head of 90,000 men, with 95 guns, he prepared to attack the French position on the Nivelle.

The French right rested on St. Jean de Luz, the centre extending along a chain of heights forming the chord of a semi-circle formed by a bend of the river Nivelle in his rear, over which were numerous bridges. On the left of this central chord was the bridge of Amots, beyond which their left stretched, having the upper part of the river in their front. The whole position was strengthened by field works, and was exceptionally strong and formidable.

Wellington, considering the French right too strong to be attacked, determined simply to hold it in check and make his attack on their centre and left.

The "Fighting Third" were, as usual, selected for some of the hottest fighting, they, together with the fourth and seventh divisions, being told off, under Beresford, to attack the French centre under Clausel. The third division were to move straight down the difficult and rocky road from Zugarramundi upon the Amotz bridge, which was defended by field works, and the possession of which would cut the communications between the French centre and left.

During the night of the 9th most of the allied troops were advanced to positions as near as possible to the French, to make the morning's work as much as possible a surprise.

As morning broke, the third division pushed on with but little opposition to the bridge of Amotz, while, on the other side of the river, Hill had sent the sixth division on the same errand. Thus Beresford and Hill drove a wedge into the French line. The third division quickly carried the bridge, and then the heights on their left, absolutely cutting apart the two wings of the French army. Thus, by eleven o'clock, Hill and Beresford were in full communication over the bridge of Amotz, holding a position in the middle of the French line, which made it everywhere indefensible. The remainder of the army pressing on, the French were soon in hasty retreat across the Nivelle; while the third division were soon pushed along the right bank of the river as far as Sante Pé, where was the chief passage across the semi-circular bend, previously described, while the French were still crossing lower down. Meanwhile Clausel had rallied some of his forces on the heights above Sante Pé, whence he was driven, after a bloody struggle, by the third and seventh divisions. The French right held on until dark, when they retired, following the rest of the army along the road to Bayonne.

Thus the first victory on French soil was won. The enemy

ATTACK OF THE
FRENCH ENTRENCHED POSITION
ON THE
NIVELLE
10th Nov.r 1813

[To face p. 120.

had lost some 3000 men, and left in the hands of the allies 1400 prisoners, 51 guns, and all their field magazines.

The allied loss was 2694 men, the 45th escaping with singular good fortune, their loss being returned as one rank and file wounded. In fact, since Vittoria its losses had been small, and no officers had lost their lives during the battles of the Pyrenees.

Soult had retired across the river Nive, and posted his army in two intrenched camps, one between the Nive and the sea, and a larger one between the Nive and the Adour.

Wellington's position stretched from Cambo, on the Nive, to St. Jean de Luz; the third division, with the 45th, were still on the right of the army, near the river.

The weather broke up to such an extent on the 11th that the country and bye roads became almost impassable, and, any further advance being for the time impracticable, Wellington was compelled to put his army into cantonments.

By the 9th of December, the state of the country once more permitting of military movements, Wellington again advanced, his left moving out to examine the approaches to Bayonne, while Beresford and Hill took the right wing across the Nive. This advance was feebly opposed, and the French were driven back towards Bayonne. Meanwhile Soult made desperate attacks, on the 11th and 12th, upon Wellington's left, but without any material success. Baffled in that direction, he attacked Hill and Beresford on the following day, but was again repulsed and driven back with heavy loss. He then concentrated his whole army round Bayonne, and the allies were advanced, occupying the left bank of the Adour.

In this position the campaign of 1813 closed, the 45th, with the rest of the third division, occupying a somewhat rearward post near Cambo.

# CHAPTER IX

THROUGHOUT January, 1814, and up to the middle of February the allied army may be said to have been drawn round Bayonne from the sea to the Adour, and throughout that period to have been preparing to advance. Many circumstances combined to delay Wellington's advance—the slowness of the Government at home, the want of clothing and necessaries, and last, but not least, the closing of the port of Santander by the Spanish authorities.

Soult's army meantime had been considerably weakened by the withdrawal of a portion of troops to reinforce those under Napoleon's immediate command.

The general position of Soult's army ran along the right bank of the Adour, with its right resting on Bayonne; the left, across the Adour, ran on the line of the Joyeuse, with its flank resting on the fortified post of St. Jean Pied de Port. The allied position extended from the sea eastwards, facing Bayonne across the Nive, and round to the Adour above Bayonne; then it turned to the south, and stretched along the line of the Nive, facing the French centre and left.

Wellington had determined to turn one of the French flanks, the left preferably, but the right if this should fail.

At St. Jean de Luz he had prepared the materials for a great bridge, which it was his intention should enter the Adour from the sea, and be completed with the assistance of the navy. He hoped, by demonstrating against the French right, to cause them to forget the lower Adour, and thus to cut off the town of Bayonne from all hope of succour; Soult being thus driven inland, and Wellington's communications with the sea kept open.

On the 14th of February Hill, on the allied right, began to advance; while Hope, concealed by woods and the windings of the river, pushed on the force which was to construct and defend the bridge. The French left was easily turned and driven back across the Bidouge.

The third division, once more under Picton, and still including the 45th, operated on Hill's left.

To avoid being driven back northward into the department of Landes, Soult retired to the Gave d'Oleron, his line lying north and south, and facing the allies on the left bank of the river; he then prolonged his line somewhat to the south, and established his headquarters at Orthéz.

On the 15th Hope made a secret advance towards the lower Adour, leaving Bayonne on his right. By the 22nd his dispositions were complete, and false attacks were made all round the intrenched camp south of the town; but the wind was treacherous, and no materials for the bridge had, in consequence, entered the river, while a heavy fire was opened on the troops from the French gunboats. But Hope was not to be denied; his artillery drove off the gunboats, and he succeeded in passing 60 of the Guards across the river in a pontoon, who quickly drove off the feeble French piquet that opposed them. A hawser was quickly passed across the river, and a flying bridge established, by means of which a considerable force was soon pushed over to the north bank.

On the 24th, after some difficulty, the flotilla, escorted by man-of-war boats, crossed the bar, and a regular bridge, amply guarded, and arranged to accommodate itself to the rise and fall of the tide, was constructed some three miles below Bayonne, which was thus practically surrounded and cut off.

Meanwhile Wellington continued to push the enemy in his front, but not without difficulty. Many of the fords across the Gave d'Oleron were found to be difficult and dangerous, and Picton's division suffered a severe repulse at Sauveterre. Higher up the river, however, he met with more success, and the sixth division crossed without opposition at Montfort, which compelled Soult to abandon the line of the Gave d'Oleron and concentrate at Orthés, on the right bank of the Gave de Pau.

On the 26th the allies began moving against Soult's position, Beresford, with the fourth and seventh divisions, crossing the Gave de Pau at Peyrehorade, and advancing on Orthéz by the Bayonne road, which ran beside the river. Later, on the same day, the third division crossed by a ford at Bereux, and, though it was not opposed, found itself at once in a somewhat difficult and dangerous position; the ground between the river and the road to Orthéz was so rocky and broken that only a few men could move abreast, and the division was within easy striking distance of Soult's left. The 45th, which was marching in rear of the division, encountered while advancing Lord Wellington reconnoitring the enemy's position in front of a small eminence. Between the brigade and the company of the 45th, which was commanded by Lieutenant Monro, was a gap of some 200 yards, which, being visible to the enemy, might have led him to think that an unsupported division alone had crossed. Wellington, observing this, came hurriedly up to Monro, and, taking him by the arm, said—"Get the other side of the hill, or the enemy will see that there are no more troops across, and be down on us!" The dispositions which followed on Wellington's directions would seem to have deceived the French, who made no attack.

The third division then took up a position which enabled Wellington to throw a pontoon bridge across the Gave de

Pau at Bereux, over which, at daybreak on the 27th, the sixth and light divisions crossed. Two-thirds of the allied forces were thus concentrated on the right bank of the river, while the remaining 12,000 under Hill, remained on the left bank, watching the bridge at Orthéz and threatening Soult's rear.

While the sixth and light divisions were struggling up from the river to the road Picton's position was almost entirely isolated. The ground to his right dipped into a plain broken up by marshes, and the nearest support on his left was Beresford's force on the hills about St. Boes. But time passed, and Soult missed his opportunity, and nothing occurred beyond some slight skirmishing between the " Fighting Third " and Foy's division, which opposed it.

The battle began about nine o'clock. The third and sixth divisions advanced easily up the slope towards the French centre and left; but the troops on the left were checked, and somewhat driven back by the French right at the village of St. Boes, and after three hours of desperate fighting the French still held their positions. Wellington now strengthened his left, and sent the 52nd regiment across the marshes to fall on the left flank and rear of the French. The 45th were deployed into line, and occupied nearly the whole of the front of the right wing of the allied army ; the Grenadier company, under Captain Martin, gallantly repelling every effort of the French to dislodge them. At no time, perhaps, during the whole war had the special characteristics of the 45th—steadiness and stubbornness— been more splendidly displayed than they were on this occasion. Under Lieut.-Colonel Greenwell, the light company steadily pushed on to the front with the 60th regiment and the light companies of the brigade, and drove the enemy slowly back. The 52nd regiment, by its flank attack, had thrown the French right into confusion, and the fourth and seventh divisions, with the cavalry, began to make considerable headway.    As the allied divisions on the left got through the village the third division flung itself impetuously on the French centre and left, which began to give way; and Soult, recognising his discomfiture, gave orders for a general retreat, leaving six guns in the hands of the allies, and over 4000 in killed, wounded, and prisoners on the field.

The allies lost heavily, the " Fighting Third " alone losing 70 officers and 800 men.    Of the 45th, Lieut.-Colonel Forbes and Lieut.-Colonel Greenwell, who led the light troops of the division were severely wounded; Lieutenants Metcalfe and Leslie were killed, and Lieutenant Macpherson so severely wounded that he never again rejoined; Captain Humphrey and Lieutenants Corby, Coglan, and Stewart were also wounded.

The story runs that just as a body of the French were about to fire a volley into the 45th, Macpherson ran to the front, waving his sword and crying, " Now, Metcalfe ! "  The two rushed against the enemy, and both fell before the same volley.

By the 2nd of March Soult, closely pressed by Hill, had retreated far up the Adour, and Wellington paused to consider his position.

In sixteen days, from the 14th of February to the 2nd of March, he had covered eighty miles, passed five large and several small rivers, forced the enemy to abandon two fortified bridge heads and many minor works, defeated him signally in one great battle, captured six guns and one thousand prisoners, forced him to abandon Bayonne, and cut him off from Bordeaux; and in all this remarkable summary of successes by no means the least distinguished of the many splendid battalions which bore the brunt of the work was the 45th.

BATTLE
OF
TOULOUSE
10th April 1814

Cavalry  Infantry  Artillery
    Allied
    French

[To face p. 127.]

After a short halt Wellington pushed on by both banks of the Adour, while Soult fell back upon Tarbes, giving up the road to Toulouse.

There were several collisions between the advanced and rear guards of the respective armies, and at Vic Bigore, on the 17th of March, the " Fighting Third " drove two French divisions from their positions; while on the 20th of March Soult, being again defeated at Tarbes, retired hastily on Toulouse, where he took up what he considered a secure position.

Wellington, whose first intention was to attack Toulouse from the south, attempted, on the 27th of March, to throw a bridge over the Garonne about six miles above the town; but the river was found to be too full and too wide, and the attempt was consequently abandoned.  On the 3rd of April, however, the third, fourth, and sixth divisions, with three brigades of cavalry—the whole under the command of Beresford—crossed the river by a bridge thrown across some fifteen miles below Toulouse.  The river then rose, injured the bridge, and prevented any support being given by the rest of the army; but Soult again failed to take advantage of his opportunity, and by the 8th of April Wellington had got most of his army over the river.

In the advance on Toulouse the third division was on the right, close to the Garonne, and consequently opposed to the strongest section of the enemy's works.  The light division was next on their left, and then Freyre's Spaniards; while Beresford, with the fourth and sixth divisions, moved along the left bank of the Ers river.  Nothing was intended to be done until Beresford should begin his attack; then Freyre's Spaniards were to make a real attack, while Picton was to make a feigned attack on the well-protected bridge of Jumeaux, which led over the canal close to the Garonne.

Two errors, however, made the earlier operations disastrous

for the allies. Freyre's Spaniards could not be held back, and, making their attack before Beresford was in position, they drew the whole attention of the French, and were in consequence driven back with great slaughter. Picton, who pushed his feigned attack too far until it became a real and desperate one, was repulsed with very heavy loss—the only occasion on which the "Fighting Third" was ever driven back, or rather obliged to be brought back, from an assault during the whole war.

The whole of the work fell on Sir Thomas Brisbane's brigade, and, as the 74th and 88th were held in reserve, the desperate and useless struggle was mainly sustained by the 45th; and it was indeed mortifying to the gallant division that its last feat in arms, after so long and brilliant a career, should have terminated in a failure, and in the loss of so many gallant officers and men killed and wounded, and amongst the latter Sir Thomas Brisbane, who had so often led on his brigade to victory.

The 45th lost its colonel, Forbes, and 7 men killed; Major Lightfoot, 1 captain, 5 lieutenants, 1 ensign, and 69 rank and file wounded.

In spite of the check to the Spaniards and the third division, Beresford's divisions carried all before them, and by four o'clock Soult abandoned the action and withdrew his whole army beyond the canal, and on the night of the 11th of April retired twenty-two miles, to Villefranche, leaving his magazines and some sixteen hundred wounded in the hands of the allies.

Hill pursued him, but uselessly, for the war was over, and on the following day messengers from Paris arrived with the news of the abdication of Napoleon and the conclusion of peace.

The 45th was mustered at Grenade, sixteen miles below Toulouse under command of Lieut.-Colonel Greenwell, and marched to Bordeaux, where it was reviewed by Lord Wellington on the 15th of June, preparatory to embarkation for Ireland.

On the 24th of June the regiment arrived at Peline, and shortly after embarked, after having served in the Peninsula for nearly six years, and won the right to carry on its colours the glorious victories of Roleia, Vimiera, Talavera, Busaco, Fuentes d'Onoro, Ciudad Rodrigo, Badajoz, Salamanca, Vittoria, Pyrenees, Nivelle, Orthez, and Toulouse.

# CHAPTER X

THE regiment landed at Monkstown on the 23rd and 24th of July, 1814, and at once marched into barracks at Cork. Lient.-Colonel Greenwell had been gazetted to the command about two months before the regiment embarked for home; he had landed at Mondego Bay scarcely six years before, fourth on the list of captains, so rapid had promotion been. In September the regiment marched to Enniskillen, in strength 44 sergeants, 22 drummers, and 694 rank and file; here a large number of discharges took place, considerably reducing the strength of the regiment; but in October the second battalion was broken up, and Captains Martin and Anderson, with Lieutenants Reynell and Barwick, were sent to Plymouth Dock to bring over what was left of it, which brought the strength of the regiment up to 54 sergeants, 22 drummers, and 836 rank and file.

Early in 1815 the corps moved to Belfast, and furnished no less than eight different detachments.

The old colours of the regiment, which had been borne all through the Peninsular War—shot to rags and shreds, and mounted on poles cut from the hedges in Spain—were now replaced by the colours of the late second battalion. There is no clear account of what became of the old colours, but it is said that they were burnt in the barrack yard at Belfast. In the following April the word " Peninsula " was ordered to be added to the honours already borne on both colours, in commemoration of the services of the regiment in Portugal, Spain, and France.

The origin of the custom, peculiar to the 45th, of carrying the honours on both colours is shrouded in mystery, but it is a distinction which the regiment has always borne.  It is said that on the outbreak of the Waterloo campaign both the 45th and 79th regiments, at Belfast, were placed under orders to proceed to Belgium, and that transports were despatched to Newry to meet them.  Colonel Douglas of the 79th, however, got his regiment first to Newry, and, embarking them, reached Belgium in time for the battle of Waterloo, while the 45th were kept waiting for further transports, which did not arrive till after the battle had been fought, and the embarkation was consequently countermanded.  During this year the head-dress of all infantry regiments, with the exception of the Grenadiers and Highlanders, was changed, and the tufted "chacko" was introduced, and remained until it was displaced by the "Albert chacko" in 1845.  In February, 1817, a detachment, consisting of 9 officers and 150 men, under Captain Martin, proceeded to Wales, returning in about a month.  Where it went is not recorded, but it is sufficient to say that it carried back to headquarters the thanks of the inhabitants of the place—transmitted by the magistrates—for the uniform good conduct of every individual attached.

Thus early did the regiment begin to establish that character for orderliness and good conduct which has ever been one of its distinguishing characteristics.

Great reductions were made in the army at this time, and in March of this year no less than ten of the lieutenants in the regiment were placed upon the half-pay list, amongst them being Sir Charles Munro.

The regiment at this period was quartered at Dundalk,

with detachments at Carrickmacross, Wickbald's Cross,
Jonesboro', Corcreagh, Louth, Ardee, and Mansfieldtown.

During the summer an epidemic fever raged about Dundalk, but care, cleanliness, and sanitary precautions brought
the regiment through with the loss of only one man.    In
1818 the regiment once more moved to Cork, where it was
placed under orders to prepare for service in Ceylon, and
on the 29th of January, 1819, the headquarters sailed in
five transports, under the command of Major Stackpoole.
The Cape was reached about the 23rd of April, and on the
9th of July the regiment disembarked in Ceylon, the headquarters landing at Trincomalee, and detachments proceeding
to Point de Galle and Colombo.    Owing to the unhealthiness
of Trincomalee, headquarters were soon moved to Colombo,
where the regiment was for a time concentrated, and subsequently moved on to Kandy.    At Colombo the regiment was
presented with new colours by Lady Brownrigg, the wife of
the Governor, Sir Robert Brownrigg, who addressed the
regiment in the following speech :—" Major Stackpoole and
officers of the 45th regiment, you have greatly honoured me
by requesting me to officiate on this occasion of replacing
your old colours, worn out in the service of your country.
Soldiers! my brave countrymen! under your old colours you
have most nobly distinguished yourselves; the new colours
you now receive are a memento of your achievements.
Thirteen victories are emblazoned upon them.    Their field
is not capacious enough to hold the rest of your deeds of
glory.    Let, then, those deeds be impressed on your
memories; let the recollection of them, let that star of
honour which shines on your hearts, lead and stimulate you
whenever and wherever duty calls you to fresh acts of
heroism.    Officers and soldiers, this island is, happily,
tranquil.    Long may it enjoy the blessing of peace; and as
on the continent of Europe you distinguished yourselves by

your valour; here may you distinguish yourselves by your virtues! This is my most sincere wish, and that you may enjoy health and every rational happiness."

During the year 1821 Lieut.-Colonel Greenwell arrived at headquarters from England, with a draft of four officers and fourteen men, and on the 27th of August resumed command of the regiment.

Nothing but a long record of sickness marks the years 1822 and 1823, except the appointment in the latter year of Captain Hamilton as A.D.C. to Sir Edward Barnes, K.C.B., the new Governor of the island. In this year also, the infantry of the line discarded the white breeches and gaiters so long worn, and adopted instead trousers of a dark grey colour. The breastplate, alluded to in the speech of Sir Thomas Brownrigg, was also discontinued.

In 1824 a small detachment of 1 sergeant and 30 rank and file proceeded from Kandy to Fort M'Dowal, where it suppressed a small native rising, and delivered the ringleaders to justice.

During this year it was found necessary, in consequence of the lawlessness of the Burmese upon the Indian frontier, to send an expedition to Burmah. The forces under Sir A. Campbell concentrated at Port Cornwallis, in the Andaman Islands, whence they proceeded to Burmah in three divisions —the main body to Rangoon, and two other columns to Cheduba Island to the north and Negrais to the west.

Before the end of the year Rangoon, the whole coast line, and all the territories accessible by water were in our hands, in spite of continuous attacks by the Burmese, which were invariably repulsed. But the climate was found to be the most difficult enemy to contend with; by September there were scarcely 1500 European troops in Burmah fit for service, 749 having died, and over 1000 being in hospital. Moreover, the scarcity of transport made operations so exceedingly difficult that nothing could be accomplished beyond the reach of the rivers.

At the opening of 1825 it was determined to make an advance inland, for which purpose reinforcements were ordered to embark, amongst which were the 45th. So thinned were the ranks of the regiment by disease that only 362 rank and file were fit for embarkation from Ceylon on the 9th of February; and it was only the imperative nature of the orders received that induced Sir Edward Barnes to permit the regiment to sail in its enfeebled state. Indeed, he thought it necessary to make a special report of the weak state of the regiment to the Commander-in-Chief in India.

Sir Edward Barnes marked his appreciation of the regiment by presenting the officers' mess with a handsome silver salver, which is still preserved, bearing the following inscription:—

"This salver was presented by His Excellency Lieut.-General Sir Edward Barnes, K.C.B., Governor of Ceylon, etc., etc., to the mess of the 45th regiment, accompanied by a note in the following words:—Sir Edward Barnes presents his compliments to Lieut.-Colonel Greenwell and the officers of the 45th regiment, and requests their acceptance of a silver waiter as a token of his respect and esteem.

"Government House,
     "Colombo, Jan. 23, 1824."

On the passage across cholera broke out, and the regiment arrived in Rangoon on the 25th of March in no condition for active service. It was accordingly at once sent away to Madras to recruit, and remained there until late in the year, thus missing all the really active and stirring work of the war.

Towards the end of the year, having been reinforced from home by a draft under Lieutenant Bell of 420 men, the

regiment again embarked for Rangoon on the transports
" Golconda " and " Earl of Kellie."

On the voyage over the " Earl of Kellie " put into the
Little Andamans for water, the watering party being covered
by a guard under the command of Lieutenant Sykes.  While
the party were at dinner they were suddenly assailed by a
shower of arrows from the surrounding jungle; one man was
killed and three others wounded, but the remainder, springing
to their arms, poured in a volley which speedily put their
assailants to flight.

On the 22nd of November the regiment landed in Rangoon,
the greater portion remaining in garrison at that place, while
a detachment of four companies marched through Pegu to
reinforce Colonel Pepper after his successful attack on
Sittang.  This detachment may probably have seen a little
service, but it is doubtful if much fighting came in its way,
as the principal seat of the war at this period was away to
the north, in the vicinity of Prome; and on the 26th of
February, 1826, peace was signed on the road to Ava.

On the conclusion of the war the regiment, together with
the 9th Madras Infantry and two squadrons of Madras
Cavalry, remained in garrison at Rangoon.

In April the Governor-General in Council issued the
following general order :—

" While the Governor-General in Council enumerates with
sentiments of unfeigned admiration the ' 13th,' ' 38th,'
' 41st,' ' 89th,' ' 47th,' ' 1st ' or ' Royals,' ' 87th,' and ' 45th '
regiments; the Hon. East India Company's Madras European regiment, the Bengal and Madras European
Artillery, as the European troops who have had the
honour of establishing the renown of the British arms in a
new and distant region, his lordship feels that higher and
more justly-merited praise cannot be bestowed on these brave
troops than that, amidst the barbarous hosts whom they

have fought and conquered, they have eminently displayed the virtues and sustained the character of the British soldier."

In the same order the Governor-General directed that the company's regiments were to bear the word "Ava" on their colours; he also stated that "with respect to the King's regiments he would recommend His Majesty, through the proper channel, to grant the same distinction to them." Medals were also ordered to be struck for distribution.

On the 3rd of August six months' batta was awarded to all troops employed for not less than twelve months in Burmah, and three months' batta to such as had served there for less than twelve months. The Court of Directors at home afterwards doubled the award.

On the 8th of February, 1827, the House of Commons voted its thanks to the troops employed in the expedition. On the 9th of November, 1826, the regiment left Rangoon for Martaban, up the Salween river, in the newly-conquered province of Tenasserim. It proceeded in transports as far as Amherst, where it was transferred to gunboats, in which it proceeded to its destination.

The chief work performed in this place by the regiment was the clearing of the jungle at Moulmein, on the opposite side of the river, as well as assisting in the erection there of what were said to be "the finest barracks in India." They took upon themselves the entire labour of clearing the ground for the cantonment, the parade ground, and the graveyard; the whole covering some 15,000 square yards. A theatre was also erected, and the whole establishment, when completed, became not only one of the best but also one of the healthiest cantonments in India. Detachments of the regiment also saw a certain amount of active service in checking the lawlessness of bands of marauders which infested the villages on the Burmese side of the river.

It was not till after six years and four months' service in

Burmah that the regiment quitted the country, embarking at Moulmein for Madras. So marked had been the steadiness and good behaviour of the regiment during its long stay in the country that petitions were sent, after the corps quitted Martaban, from the Burmese begging for its return.

On the 14th of April, 1832, the corps was inspected at Poonamallee, preparatory to marching for Arnee, by the commander-in-chief, Lieut.-General the Hon. Sir R. W. O'Callaghan. He praised the regiment highly, declaring to Colonel Shaw that he ought to be extremely proud to command such a regiment, and desired that his sentiments might be made known to the officers and men in an order of the day.

The regiment, which numbered 44 sergeants, 13 drummers, and 799 rank and file, arrived at Arnee on the 26th of April, and on the 28th had to lament the death of Captain Perham. It was intended that the regiment should continue its march to Secunderabad in September, but a severe attack of cholera occurring in the district, which attacked the regiment and carried off no less than 109 men, caused the order to be changed, and the regiment moved off instead to Bellary, and ultimately continued its march to Masulipatam, on the west coast, where it remained for some months. It was relieved here by the 62nd regiment on the 15th of April, 1833, on which date it marched for Secunderabad.

During its stay at Secunderabad fever raged in the regiment, and in the spring of 1834 Lieutenant Armstrong and Lieutenant Rose both fell victims to disease. So great was the sickness in October that the regiment was moved out under canvas with a view to checking it. Notwithstanding these adverse circumstances the character of the regiment still stood as high as ever, and Lord Hill, the commander-in-chief in Madras, wrote as follows on the 3rd of August,

1834 :—" The General commanding-in-chief did not fail to observe that courts-martial in the regiment since the previous inspection were very limited in number; and, moreover, that no case of corporal punishment had occurred. This statement has been very satisfactory to Lord Hill, who considers it very creditable to the commanding officer and other officers of the regiment, as evidencing on their part a degree of zeal in the discharge of their respective duties by which the above essential object has no doubt been mainly attained."

On the 21st of April, 1835, Major Poyntz died, and Captain Thomas Ewan obtained the vacant company; he fell ill, obtained sick leave, and died at Godoor on his way home; and on the 22nd of May, in the same year, Lieutenant Moore died.

On the 18th of November, 1836, the 45th marched out of the cantonments at Secunderabad to make room for the 55th regiment, and commenced the first march on their homeward journey, after a period of foreign service almost unparalleled for sickness, as the following inscription from the regimental memorial, subsequently erected at Secunderabad, will show :—

"Erected to the memory of 22 officers, 70 sergeants, 44 corporals, 17 drummers, 995 privates, 163 women, and 183 children of the 45th or 1st Notts regiment, who have died from the date of embarkation to India, January, 1819, till the 18th of November, 1836, when the regiment marched for Arnee, preparatory to its return to England.

"Died at Secunderabad, 4 officers, 12 sergeants, 8 corporals, 3 drummers, 116 privates, and 22 women."

The resident at Hyderabad, Colonel Stewart, marked the

[To face p. 138.

occasion by the following letter to Lieut.-Colonel Freeman, commanding the Hyderabad subsidiary force :—

" Sir,

"On the occasion of the departure of H.M.'s 45th regiment from the Nizam's dominions I consider it my duty to state that, during the whole time that the regiment has been stationed here, not a single instance of any complaints against any man of the regiment by any of the Nizam's subjects has ever been brought to my notice. Such a strong proof of the good conduct of the corps I conceive demands the public expression of my acknowledgment, and I request that you will convey to Lieut.-Colonel Boys and the officers and men of that distinguished regiment these my sentiments, together with my sincere wishes for their future welfare."

The following order was also issued about the same time :—

"Headquarters,
"Hyderabad Subsidiary Force.

"Divisional Morning Orders.    By Lieut.-Colonel Freeman.

"Secunderabad,
"Monday, 21st November, 1836.

"His Majesty's 45th regiment having quitted Secunderabad, and encamped preparatory to its return to England, Lieut.-Colonel Freeman performs a pleasing duty in recording the orderly, praiseworthy, and exemplary conduct of this regiment during the period of its service with the Hyderabad subsidiary force. It is a theme of general admiration, and this excellent regiment justly merits and receives the tribute of respect and regret naturally consequent on its separation from this cantonment. After a long period of service in India, with a reputation established for discipline and good conduct in quarters, and its banners covered with honours which it has gained in war, H.M.'s 45th regiment has secured earnest hopes for its safe return to our native land, and for its future welfare and success. Assured that a happy union of sincere and right good will to H.M.'s 45th regiment.

"By order.

(Signed)    "H. S. FAYNE,

"Assistant Ad.-General."

On the 21st of January, 1837, the regiment arrived at its old quarters at Poonamallee, and the headquarters embarked at Madras for England on the 18th of the following November, and, after touching at Cape Town, landed at Gravesend on the 23rd and 24th of March, 1838, after an absence in the East of nineteen years and two months. It was naturally a changed regiment after so long a service abroad, and of the 800 men who had embarked at Cork nineteen years before but 22 returned with the regiment.

# CHAPTER XI

1838 to 1843—Thorne riot at Canterbury—Riots at Newport—Reserve
Battalion formed—Embarkation for the Cape.

ON its arrival from the East the regiment proceeded to
Canterbury, where it was joined by the depôt from Chatham.
It was during its stay at Canterbury that the affair at
Bossenden Wood took place, in which one officer and a few
men lost their lives.

A few years before a Cornishman of the name of John
Nicholls Thorne, of remarkably fine presence, and with a
remarkable gift of natural eloquence, but of humble birth
and of weak moral character, left his home in Cornwall and
made his appearance in Kent, where he claimed to be Sir
William Courtenay, a Knight of Malta, and to have been
unjustly deprived of large estates in the county. He
dressed himself with extravagant magnificence, and went
from village to village in the neighbourhood of Canterbury
addressing the farmers and rustics in wild and extravagant
speeches. In 1833 he became a candidate for the representa-
tion of Canterbury, and actually polled 960 votes. But not
long afterwards he was concerned in certain transactions
which brought him within the grip of the law, and he was
sentenced to six years' transportation. However, as he was
considered to be mad, he was removed from Maidstone gaol
to the county lunatic asylum, where he remained for four
years. His parents then succeeded in bringing pressure to
bear on the Home Secretary, Lord John Russell, who foolishly
caused him to be released, and in the spring of 1838 Thorne

re-appeared in his old haunts, but added to his former claims that of being the new Saviour of the world.

He took up his abode in the village of Boughton, about six miles from Canterbury, and acquired considerable influence on the score of his Messianic pretensions, not only among the ignorant peasants but also among the more respectable farmers, who might have been expected to know better. As his madness grew and developed so did the extent of his following. He gave out that he was invulnerable, proposed to work miracles, and assured his dupes that in following him they would become invulnerable also. On the 28th of May, 1838, with about 100 followers, partly armed, he started from the village of Boughton and visited the villages of Goodnestone, Hernehill, and Dargate Common, where he divested himself of his shoes and proclaimed that he now stood on his own holy ground, while his deluded followers fell on their knees around him.

That night the fanatics slept in a barn in Bossenden Wood, and the next day visited the villages of Newnham, Eastling, Throwley, Selwich, Lees, and Selling, a circuit of some seventeen miles, and sleeping, presumably, at the latter place. On the 30th of May they returned to Bossenden Wood. No breach of the peace had occurred up to this date, but on this day a farmer, angry at the withdrawal of his labourers, made an application for their apprehension. Three constables were despatched on the morning of the 31st of May to carry out this duty; they had no difficulty in finding Thorne, who, after a short parley, inquired who was the leader of the party. On being informed, he drew a pistol and shot him dead; then hacking at his body with a sword which he carried, exclaimed to his followers, "Now, am I not your Saviour?"

The two uninjured constables made off as rapidly as possible, and reported what had occurred to the magistrates. The matter had now become serious; word was sent back to

Canterbury, and 100 men of the 45th, under Major Armstrong and Lieutenant Bennett, were sent at once. The rioters retreated to a deep and sequestered part of Bossenden Wood, where Thorne harangued his followers and excited them to desperate fury.

The detachment of the 45th was divided into two parties, one, under Lieutenant Bennett, entering the wood. They had no difficulty in finding the fanatics, and Lieutenant Bennett, advancing in front of his men, summoned Thorne to surrender. The madman came forward to meet him, and suddenly, before anyone was aware of his intention, shot the officer dead. The men of the 45th immediately fired a volley in return, and Thorne fell mortally wounded, exclaiming, "I have Jesus in my heart!" His followers rushed forward, and for a few moments sustained a hand-to-hand conflict with the troops. In a short time, however, the rioters were dispersed with the loss of 8 killed and 7 wounded, two mortally. The 45th, besides Lieutenant Bennett, lost two killed and one wounded. Inquests were held on the bodies of the slain, and verdicts of "wilful murder" were returned against Thorne and eighteen of his followers, while the act of the troops in shooting down the rioters was pronounced to be "justifiable homicide." Lieutenant Bennett was buried with military honours in the cloisters of Canterbury Cathedral, and a tablet was erected to his memory, with the following inscription :—

"Within the cloisters of the Cathedral are deposited the remains of Henry Boswell Bennett, lieutenant in the 45th regiment, who fell in the strict and manly discharge of his duty in Bossenden Wood, in the ville of Dunkirk, on the 31st of May, 1838, aged 29 years. As a lasting mark of sincere regret for the melancholy loss of an amiable and esteemed companion, this tablet is erected by his brother officers."

In March, 1839, the regiment took its first voyage by steamer, embarking at Herne Bay for London, *en route* to Windsor, which was reached by rail from Paddington to Slough, whence the regiment marched to its destination, a detachment having been previously sent to the Royal Palace at Kew. During the stay of the regiment at Windsor it received new colours, but they were not, for some unknown reason, formally presented. These identical colours, barring some considerable repairs, are still carried by the regiment, and are undoubtedly the oldest in the British army. In October the regiment marched to Winchester, after detaching Captain Mellor's company to Newbridge and Captain Stack's company to Newport, Monmouthshire.

The detachment sent to Newport was called for in consequence of the disturbed state of the country owing to the Chartist agitation, which, with Birmingham for its headquarters, had strongly infected the neighbourhood of Newport. The Chartist leader of the district was a linen draper named Frost, who had unfortunately been placed on the bench of magistrates; but, having proved himself totally unfitted for the position, he was called to account by the Home Secretary, Lord John Russell. Frost defied him, and his name was accordingly early in 1839 struck off the Commission of the Peace, with the result that he assumed the leadership of the Chartists in the district.

Just before the arrival of the detachment Frost and his colleagues—Zephania Williams, a beershop keeper, and William Jones, a watchmaker—had mustered their forces, and arranged a scheme whereby they were to arm the Chartists of the surrounding districts and seize and hold Newport as a Chartist centre. It was expected by them that this would be a signal for the rising of the Birmingham Chartists, to be followed by a general rising all over the north of England. There is but little doubt that Frost mustered some 15,000 men, to repel which there was but the single company of the 45th just arrived in Newport.

The night fixed upon for the attack was Sunday, the 3rd of November, and the rioters proceeded to assemble in three bodies at three different points, under the three leaders before-mentioned, Frost, Williams, and Jones, and to concentrate at one point, the Welsh Harp, and thence to march upon the town in time to begin the attack not later than one o'clock on the Monday morning.

Rain fell in torrents. Frost's division, numbering 5000 men, was the first at the place of rendezvous, and, both the other divisions being delayed by the weather and the length of the march, he advanced at the break of day without them, through Tredegar Park upon Newport.

The mayor, Mr. Thomas Phillips, having notice of what was coming, had sworn in special constables, who occupied the three principal inns of the town; while the mayor and the magistrates had their headquarters at the Westgate Hotel, in the Market Place.

Captain Stack had taken up his position, with his detachment, in the poorhouse; but about eight o'clock, in response to an urgent request from the mayor, he sent Lieutenant Basil Gray, with 30 men, to defend the Westgate Hotel. The hotel faced the Market Place, and was entered by a courtyard with a gate that could be closed. There were two large rooms, one at the east and the other at the west end of the house, connected by a long corridor. At the request of the mayor Lieutenant Gray marched his men into the house, and the gate of the courtyard was shut; he then took up his position in one of the large rooms which had three projecting windows, which were fitted with half-shutters; the furniture was cleared from the room, and the shutters were closed to conceal the men.

These preparations were scarcely completed when the
K

cheers of the approaching mob were heard.    Frost had divided his force into two parties, with orders to enter the town from two different directions; one he commanded himself, while the leadership of the other was entrusted to his son, a boy of fourteen or fifteen.

Aware of the presence of the troops in the hotel, Frost arranged for the two bodies to concentrate in front of the building; which manœuvre being duly completed, he summoned the special constables, whom the mayor had placed to guard the entrance of the hotel, to surrender.    The mob quickly bore down their resistance, and fired down the passage into the bar and into the windows of the room where the troops were concentrated.    Lieutenant Gray now gave his men orders to load, but during the time occupied in performing that operation some of the mob had pushed through into the bar.    Gray and the mayor now opened the shutters; the rioters at once discharged a volley through the windows by which the mayor and Sergeant Daly were wounded.    But the triumph of the mob was short-lived.    The windows enfiladed the street, and the 45th detachment were quickly at work, and in ten minutes the rioters had fled in every direction, leaving 20 killed and a great number of wounded inside the hotel and in the Market Place.

Two remarkable facts with regard to this little party of the 45th deserve to be mentioned.    It was the second time it had been employed in quelling civil disturbances, for it was Captain Stack's company that had dispersed the rioters in Bossenden Wood in the previous year; and, secondly, the extreme youth of the party engaged, there being only three men in the detachment who were over 23 years of age. Captain Stack was not molested in his post at the poorhouse, the dispersal of the mob being absolutely accomplished by the skilful conduct of Lieutenant Gray; but all concerned received the commendation which was due to them for steadiness and gallantry which nipped in the bud what might otherwise have become a grave national danger.

A public meeting was held at Newport, at which resolutions were passed awarding "the warmest thanks" of the community to "Captain Stack, Lieutenant Gray, Ensign Stack, and the non-commissioned officers and men of Captain Stack's company of the 45th regiment." It was also resolved "that a humble memorial be presented to Her Majesty expressive of the gallantry, coolness, and intrepidity of Captain Stack, Lieutenant Gray, Ensign Stack, Sergeant Daly, and the non-commissioned officers and privates of Her Majesty's 45th regiment of foot stationed at Newport on Monday, the 4th of November, respectfully soliciting Her Majesty graciously to bestow some substantial mark of gracious favour and distinction upon the officers, non-commissioned officers, and privates before named." As a result Captain Stack received his majority and Lieutenant Gray was promoted to an unattached captaincy.

On the night of the 6th of November a special messenger arrived at headquarters with orders for the regiment to proceed to Newport by forced marches. It started early the next morning, and marched into the town on the 10th, where it was subsequently reinforced by a battery of Horse Artillery and a troop of the 10th Hussars. Before they arrived, however, the disturbances were at an end and the leaders all prisoners. They were tried by a special commission, and received due punishment, Frost being sentenced to death; but the sentence was subsequently commuted to penal servitude for life.

The regiment was now broken up into detachments in various parts of Wales—at Brecon, Merthyr, Pontypool, Swansea, Newtown, Llanidloes, and Montgomery, where it remained until once more concentrated at Cardiff, in October, 1840, in readiness to embark for Ireland.

During the period that the regiment was quartered in Wales Major Stack died from a chill caught by exposure, as well as Major Armstrong, who died at Monmouth in June, 1840. In October the regiment sailed in detachments from Cardiff for Belfast. While at Belfast, Lieut.-Colonel Boys caused the standing orders of the regiment, which had been in existence since 1741, to be collected and printed, and they appeared on the 1st of January, 1841, with the following introduction prefixed:—

"1. The advantage arising from a uniform and undeviating system which points out to all ranks the more prominent duties required from each are so obvious that these regulations have been compiled from various well-matured sources with a view to establish in the 45th habits of regularity, discipline, and subordination, without which a corps can neither be respectable in its own eyes, useful to its country, or formidable to its enemies.

"2. Lieut.-Colonel Boys desires that henceforward they may be considered as the standing orders of the regiment, subject to such alterations and additions as the circumstances of the times, the orders of the commander-in-chief, or the general officers under whom the regiment may serve, may make requisite. But upon no account are they to be deviated from without such necessity, which will always be notified to the corps in an official manner through the customary channel.

"3. In the hope, therefore, of inducing the officers generally to dwell upon the importance of the duties which each and all have to perform, the Lieut.-Colonel has arranged the following brief abstract under separate heads for the more easy observation of the several members of the corps, signifying the general duties of every department. The economy of companies is particularly dwelt upon; as a good system, regularly and uniformly established in the different companies, and strictly enforced, must considerably contribute to the efficiency of the regiment.

"4. The commanding officer takes the opportunity of enforcing on the minds of those in authority that the soldier should be taught to look up to his officer as a patron and protector; and it becomes the particular duty of every officer to regulate his conduct so as to afford a constant example of correctness to those under his command.

"5. Attentive to their own dress and conduct, they are to suffer no impropriety in those of the non-commissioned officers and privates, but are to scrutinise the appearance of the men when off duty as well as when on duty, considering themselves called upon on all occasions to instruct the ignorant as well as to reprimand those who persevere in irregularity.

"6. Officers cannot too often inspect the situation of men under their command; they should study their wants and, so far as circumstances will permit, promote their comfort.

"7. All officers and non-commissioned officers are directed to be in possession of a copy of these orders, and, with a view to ensure strict conformity, copies of reports and forms are herewith attached.

"C. A. EDMUND FRENCH BOYS,
"Lieut.-Col. Commanding 45th Regiment.
"Infantry Barracks, Belfast."

In August, 1841, the regiment moved to Dublin, and in April, 1842, its strength was augmented, and the establishment fixed as follows:—1 colonel, 1 lieut.-colonel, 2 majors, 12 captains, 14 lieutenants, 10 ensigns, 1 paymaster, 1 adjutant, 1 quartermaster, 1 surgeon, 2 assistant surgeons, 48 sergeants, 1 drum-major, 24 drummers, 1140 rank and file, with 60 corporals, 1 sergeant-major, 1 quartermaster-sergeant, 1 paymaster-sergeant, 1 armourer-sergeant, 1 schoolmaster, 1 orderly-room sergeant, and 12 colour-sergeants.

In September, 1842, the regiment left Dublin by sea for the Mediterranean. On reaching Cork, however, the plans were changed, and instead of proceeding it disembarked and went into garrison there.

The regiment was now divided into two battalions of six companies each; the first battalion was commanded by the lieut.-colonel, and the second, which was styled the reserve battalion, was commanded by one of the majors. On the 10th of October the reserve battalion, under the command of Major Butler, proceeded to Kinsale, but rejoined the first battalion at Cork two months later, where the whole was inspected and reported fit for service anywhere by Sir Octavius Carey.

On the 31st of January, 1843, orders were received for the regiment to hold itself in readiness to proceed to the Cape, and in consequence the reserve battalion was again separated from the headquarter battalion, remaining in Cork and furnishing detachments to Buttevant, Ballincollig, and Dungarvan. During the course of the year two companies, under Major Erskine, were employed in aid of the civil power at Bandon and Skibbereen on the occasion of one of O'Connell's monster meetings.

On the 24th of February the headquarter battalion sailed for the Cape on the "Thunderer," with the exception of one company under Captain Shaw, with Ensign Kipper and Surgeon Menzies, which sailed on the 28th in the "Radamanthus" and the "Rodney," which also conveyed the 7th Dragoon Guards. That steadiness and reliability which had long characterised the regiment was again evidenced on their embarkation, in spite of the fact that it was full of young soldiers—the average age being only twenty-three. "The battalion," wrote Major Butler, commanding the reserve battalion, to the colonel of the regiment, Sir J. G. Graham, Bart., "marched from the barracks at half-past ten o'clock this morning, not a man being absent or irregular in any way. The major-general was highly gratified at their soldierlike and steady appearance." At this time there existed in the regiment a strong feeling that in distinctive titles it had not received all that its services merited. The militia corps of Nottingham bore the title of "Sherwood Foresters," which was coveted by the regiment. Moreover, there was some strong feeling in the matter, as the title had been asked for by the 59th, or 2nd Nottinghamshire regiment, which had caused Lord Hill to express a strong opinion that, if conferred at all, the title belonged to the 45th.

Before embarking for the Cape Colonel Boys had forwarded an application to the colonel of the regiment, begging him to lay before the Duke of Wellington the request that the regiment should receive the title of "Royal Sherwood Foresters." These letters appear to have gone forward, though nothing at the time came of the application.

# CHAPTER XII

THE "Thunderer" touched at Santiago, Teneriffe, on the 17th of March, 1843, for water, sailing again on the 21st, and arrived at Simon's Bay on the 30th of April after a pleasant but uneventful voyage. The regiment landed on the 3rd of May, and marched the same day 25 miles to Cape Town; a detachment was left at Simon's Town, and another sent to the convict station at Robben Island. Affairs at the Cape were, at this time, in a somewhat disturbed condition. The Dutch colonists, who were gradually working their way further up country in the hope of escaping from a government which they considered unjust, were being brought into closer contact with the native races, with whom they had never been on good terms; while the natives, on their side, availing themselves of the feeling in their favour on the part of the Government, claimed equal rights, which specially irritated the Dutch and fostered the spirit of war.

This pressure had had its greatest effect between 1815 and 1835, when the Hottentot question, the slave question, and the Kaffir question gave the greatest fillip to the wandering tendencies of the Dutch farmers, and sent them further afield. After considerable quarrelling and bloodshed, on the 12th of May, 1843—that is nine days after the landing of the 45th—Sir George Napier, the Governor of Cape Colony, issued a proclamation constituting the district of Natal a British colony, and appointing Mr. Henry Cloete the first commissioner in the government of the new territory.

This caused further confusion among the Burghers, who had settled in Natal to escape what they considered the grievances of British law, when they again found themselves brought under the law they had hoped to rid themselves of. In addition to this, Kaffir raids across the Great Fish River—the eastern boundary of Cape Colony, in which the cattle were driven off and their owners murdered—were of almost daily occurrence.

It was the general opinion that the line of the Fish River was indefensible against these raids, and that a line some twenty-five miles in advance should be adopted.    These raids were at their worst at the time the regiment landed in South Africa, and it had already been decided to make reprisals on a large scale.    Colonel John Hare was at this time Lieutenant-Governor of the Eastern District, and responsible for the defence of the frontier; and the troops available consisted of four battalions of infantry, three of which, the 27th and two battalions of the 91st, were on the frontier, the 45th in Cape Town, the 7th Dragoon Guards, the Cape Mounted Rifles, and some artillery and engineers. On the 15th of July the detached company at Simon's Town rejoined headquarters; and about the same time two companies under command of Captain Hind, with Captain Kyle, Lieutenants Blenkinsopp and Armstrong, Ensigns Miller and Egginton, and Assistant-Surgeon Best, were detached to Natal, where they arrived on the 22nd of July to assist in bringing the Burghers under subjection and establishing the Queen's authority in the new colony.    Here they placed themselves under the orders of Brevet-Major Smith, who was already in command, with a detachment of the 27th regiment.    The chief difficulty with the Dutch farmers now arose from their ignorant belief that the Netherlands Government would take them under its protection. That Government had, of course, repudiated in the strongest terms any intention of interfering in the affairs of a friendly Power, but a paper circulated in Natal to inform them of the fact was held by the Burghers to be spurious.

Before the arrival of Mr. Cloete on the 5th of June, the Burgher Government of Natal had been *in extremis*. The closing of the port had destroyed its revenue—there was not sixpence in the treasury, and even the meagre salaries of the officials were months in arrear.

Just about the same time Panda, the chief of the Zulus, in an access of ferocity, had embarked on such a career of massacre that his subjects fled in thousands into Natal to seek the protection of the white man. The Burghers, in terror at the influx of some fifty thousand Zulus, besought Major Smith to drive them back, which he refused to do; and instead informed Panda of his intention to protect the fugitives.

By the month of July quarrels among the leading Dutchmen at Pietermaritzburg had induced a majority to give in their submission to the Queen's Government; this was accepted by Commissioner Cloete, who sent at the same time for a force to occupy the capital. Accordingly, on the 25th of August, Major Smith, taking with him the detachment of the 45th and detachments of other regiments—in all about 200 men—marched for Pietermaritzburg, where he arrived on the 31st, and at once took post on a hill overlooking the west end of the town.

One of the officers of the 45th has left on record the delightful experiences of the party at escaping from the routine of Cape Town to the pleasant novelties of country and circumstance in Natal. The climate was described as delightful, and the " wattle and daub " huts which had been built by the 27th regiment were found to be so much more comfortable than they looked. The five days' march through a lovely country was made more exciting by the expectation of attack the whole way. The author speaks in admiration of the skill of the engineer officer who accompanied the little column, "who, though he had never seen the place before (all his information being furnished by a plan made by someone else), yet at once chose the most defensible position, and proceeded to occupy it as if the whole country were well known to him." The guns were so posted as to command the town, and the officers and men, after providing a sufficient guard against surprise, were told off into working parties to construct permanent habitations and the necessary works for the protection of the post.

At first there were many alarms of night attacks, but the Burghers had neither the spirit nor the means to carry them out. The Dutch for a time were sullen and uncivil in their demeanour, but as they mixed more with the troops this wore off; the traditional capacity of the 45th for making themselves liked produced its usual effect, and as months rolled on the soldiers and the Dutch settled down peaceably together on terms of friendship.

In May, 1844, the new Governor of Cape Colony, Sir Peregrine Maitland, K.C.B., visited Pietermaritzburg, where he inspected the detachment of the regiment. He was an old friend of the 45th, having known it in the Peninsula; and it had also served under him in 1836-7, when he was commander-in-chief in Madras. To both of these circumstances he made allusion when expressing his pleasure at meeting the regiment again, and finding its character so well maintained by Captain Hind's detachment.

On the 2nd of May, 1845, the light company, under Captain Charles Seagram, was detached from headquarters to reinforce the troops on the eastern frontier. It embarked on the "Thunderbolt" for Port Elizabeth, where it arrived on the 7th of May, and at once marched to Grahamstown, which was reached on the 11th, and two days later crossed the

Great Fish River to Victoria Post, then under the command
of Colonel Somerset. On the 25th of July it was again on
the move, being ordered to Colesburg. Before the company
quitted his command Colonel Somerset issued an order
expressing his "entire approbation of the steady conduct and
soldierlike bearing of the detachment during the period it
had served under his orders." The march to Colesburg was
over 200 miles across a country almost roadless and embrac-
ing two ranges of hills. The company arrived at its destination on the 17th of August, and was quartered in storehouses
rented by the Government.

On the 24th of July the headquarters at Cape Town
received orders to embark for Natal, and marched on the 1st
of August, in torrents of rain, for Simon's Bay, where they
embarked on the "Thunderbolt." The disembarkation in
surf boats at Port Durban was scarcely a novelty to a
regiment that had twice landed at Madras, and was completed without accident. The regiment took up its quarters
in temporary barracks at Durban, two miles up the river.

Lieutenant-Colonel Boys, being the senior officer present,
was appointed commandant of the district; he also became
*ex-officio* member of the Council and acting-governor in
the governor's absence. The headquarters of the force in
Natal was established in Pietermaritzburg, whither Lieut.-
Colonel Boys proceeded with the companies under his
command, the two companies which had previously been
quartered there going down to garrison Durban. On arrival
at Pietermaritzburg the headquarters found substantial stone
barracks ready for them, capable of holding 200 men, which
had been built by the detachment, and which were noted for
years afterwards as being the most comfortable in the colony.
Fort Napier, as the post was now called, was a defensible
barrack, quadrangular in form, and situated on an eminence

to the west of the town. The outer walls were windowless and loopholed, the windows being on the inner side, looking into the square, which was 200 feet long. At the east and west angle of the barracks there were redoubts, mounting three guns on revolving platforms, and completely commanding the town.

All was now quiet in the neighbourhood of Pietermartzburg, and the garrison was mainly employed in enlarging the barracks. A garrison chapel was erected, in which the first service was held on the 14th of November, 1846; a theatre was also built, in which the first performance, before the lieut.-governor, took place on the 3rd of March in the same year. The names of Lieutenant Gibb and Lieutenant Armstrong, of the 45th, are specially connected with these excellent examples of what soldiers can do when required.

On the 24th of March there was a great military display in honour of the reception of the chief of the Zulus, when the 45th and a detachment of the Cape Mounted Rifles paraded together, but otherwise hunting expeditions were the only break in the ordinary routine of life. Game appears to have been plentiful in the neighbourhood in those days, as the following recorded bag testifies, viz., one elephant and twenty-six eland, while ten lions and two more elephants were seen but not killed. Changes in the composition of the reserve battalion in this year promoted Major Archibald Erskine, of the 1st battalion, to the command of the reserve battalion, and gave the vacant majority to Captain Hind, commanding the detachment at Durban; and during the year Ensign Morley died at Port Natal.

We must now return for a while to chronicle the history of the light company, which we left on the 17th of August, 1845, at Colesburg.

The Dutch farmers of the north-eastern districts of Cape

Colony had been long accustomed, in seasons when the grass failed on their farms, to drive their cattle up to and over the Orange River, returning when the drought had come to an end. Gradually the frontier of the colony became extended in this way, and when the way was found there by the emigrant farmers who desired to escape from British law, as well as those who had quitted Natal, there grew up a considerable population on both sides of the Orange River of Dutch farmers, as well as of the native races, who lived in a very unsettled state of government.

There were, in fact, several sections of the native population, and at least two sections of the Dutch farmers, hostile to one another, and ready, if permitted, to put their pretensions to the test of war. The Cape Government had not sufficient forces to preserve order, and yet it was evident that some display was necessary. Accordingly, in 1842, portions of the 91st regiment, 27th regiment, and the Cape Mounted Rifles, with some artillery and two guns, had been thrown forward to Colesburg under Colonel Hare, but were subsequently withdrawn, leaving only two companies of infantry under Captain Campbell, and a troop of the Cape Mounted Rifles under Captain Donovan, in camp at Colesburg.

In April, 1845, Captain (now Major) Campbell had found it necessary to cross the Orange River with his small force, and to march to Philippolis to interpose between the factions and keep the peace; and on the 26th of April his force was reinforced by the arrival of the 7th Dragoon Guards and a troop of the Cape Mounted Rifles under Colonel Richardson. There was a skirmish with the hostile Dutch farmers, who, taken in rear, submitted after very slight opposition. But a continual display of force was necessary to keep the peace, for which purpose the light company of the 45th had been despatched to Colesburg. Captain Seagram, upon arrival,

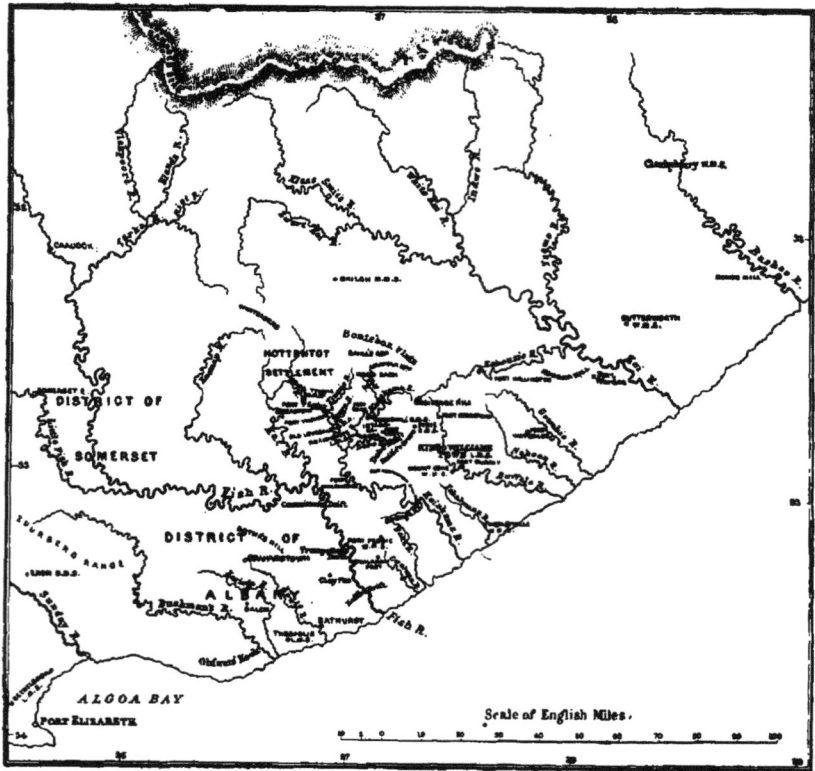

Scale of English Miles.

[*To face p.* 159.

proceeded to fortify his post, and mounted some guns which he found in store, training his men to the use of them.

In the opening months of 1846 events began to develop. A Kaffir stole an axe at Fort Beaufort, and was sent under escort to Grahamstown for trial. On the way Kaffirs attacked the escort, killing one man, and rescued the prisoner. Colonel Hare at once demanded the surrender of the prisoner and the murderers of the escort from the chief Tola. This was refused, and Tola was supported in his refusal by Sandile, the paramount chief. Colonel Hare was accordingly compelled to resort to arms. The settlers were armed, and the garrisons of the frontier posts, Forts Peddie and Beaufort, were strengthened, and preparations made for an advance on Sandile's kraal. The Kaffirs began operations by plundering all the traders in their vicinity, and on the 31st of March Colonel Hare issued a proclamation calling all the Burghers to arms. Sir Peregrine Maitland at once sent forward all the troops that could be spared from Cape Town, consisting of 80 men of the 27th regiment under Lieut.-Colonel Johnstone, and two guns under Captain Eardley Wilmot, R.A., and followed himself on board H.M.S. "President" on the 1st of April.

The force on the frontier was quite insufficient to meet the necessities of the case; it consisted of detachments drawn from the two battalions of the 91st regiment, in all 994 men; the 7th Dragoon Guards, 330 men; some 400 of the Cape Mounted Rifles, and a few Royal Artillery and Royal Engineers. The greater part of the infantry were required to guard the frontier posts; and as an addition to the field force, 1500 of the Hottentot settlers, named "Stockenstrom's Hottentots," on the Kat Kat River, were called out.

Colonel Hare, believing that a blow struck immediately

would bring about Sandile's submission, made an advance
on his kraal, which was situated about a day's ride from the
advanced frontier post called Fort Victoria.  The kraal was
found deserted, Sandile having retired to the stronghold of
Amatola, whither he was followed by a column under Colonel
Somerset.  The column, however, was caught in an ambush,
suffered heavily, and had to retire with the loss of all its
waggons.  This success at once roused all the Gaika tribes,
who poured over the border, bent on murder and destruction.
The settlers defended themselves in laagers as best they
could, but there was terrible loss of life and property.  The
outposts on the Fish River were barely strong enough to
defend their posts, and were powerless to check the general
raid.  Sir Peregrine Maitland, who had arrived at the front,
assumed command of all the forces, and proclaimed martial
law on the 22nd of April.

On the 26th of April an express arrived at Colesburg
ordering the detachment of the 45th to march with all speed
to Fort Beaufort.  Accordingly it quitted that place on the
29th, leaving behind it the goodwill and respect of all the
inhabitants, as the following letter from Mr. Rawston, the
civil commissioner, shows : —

                              " Colesburg, 18th May, 1846.
     " Sir,
          " We, the undersigned inhabitants of the village of
Colesburg, request that you will allow them to forward to
you their testimony of the good conduct of the detachment
of the 45th regiment, lately stationed here under your
command.  Their correct military deportment as soldiers,
and general orderly and peaceful behaviour as citizens, have
alike been the subject of our remark and approval.  Although
the circumstances of the Colony now ask for their presence
elsewhere, they do not, we trust, forbid us the hope that in

more peaceful times they may return to re-occupy their late quarters.    Meanwhile our best wishes will attend them, including your brother officers and yourself.

"We have, etc.,

"L. J. RAWSTON,

"Civil Commissioner

"(and others).

"To Captain Seagram,
"45th Regiment."

The local newspaper at the same time wrote:—"Great credit is due to the commanding officer of the detachment, who, well knowing the duty that called his services to this part, has, by his conciliatory line of conduct and excellent discipline, quite overcome that prejudice which naturally exists, especially in this district, between the civilians and the military.    Not one instance of civil rights has he outraged.    The greatest deference was paid in his protection of such, and the detachment of the 45th which marched to-day carried with it the good wishes of all classes."

The company was met on its route by counter-orders directing it to halt at the Tarka River, sixty miles short of Fort Beaufort, and to adopt every means to strengthen its post.    It reached its destination on the 8th of May, and proceeded to carry out its orders with such skill and vigour that six weeks later Sir Andrew Stockenstrom, after inspecting the works, reported to the governor that it was one of the strongest posts in the district.    Colonel Shore, on receiving the report, wrote as follows:—

"Fort Beaufort, 3rd July, 1846.

"Sir,

"I request you will accept my best thanks for the sketch of a fort which has been erected at Tarka Post by the

L

detachment of the 45th under your command. The work described is very creditable to yourself and the excellent company which you have the good fortune to command. I request you to convey to your officers and men the high sense entertained by the commander-in-chief and myself of their good service and very exemplary conduct since their arrival in this command.

"I have, etc.,
(Signed)     "J. HARE, Colonel.

"To Capt. Seagram,
"Light Company, 45th Regiment,
"Tarka Post."

Captain Seagram and his company, about 80 strong, remained in occupation of the fort they had erected until the beginning of August.

After his reverse Colonel Somerset retired within the frontier, leaving Fort Peddie invested by the Kaffirs. A convoy of supplies was despatched thither from Grahamstown on the 18th of May, under an escort of 80 men of the 91st regiment and 40 Burghers, but was captured by the Kaffirs, and the condition of Fort Peddie became critical. It was, however, relieved by Colonel Somerset, with a force of 1200 men, after defeating the Kaffirs at the Fish River.

Following up this success, Colonel Somerset attacked and defeated with great slaughter a body of the Kaffirs which had formed an ambush to surprise him on the 7th of June, and by the end of the month a considerable force had assembled on the frontier, composed as follows :—

| | |
|---|---:|
| 7th Dragoon Guards - - - - | 325 |
| Royal Artillery - - - - | 114 |
| Royal Engineers - - - - | 155 |
| Two battalions 91st Regiment - - | 983 |

| | | | | | |
|---|---|---|---|---|---|
| 27th Regiment - | - | - | - | - | 416 |
| 45th Regiment - | - | - | - | - | 151 |
| 96th Regiment - | - | - | - | - | 439 |
| Cape Mounted Rifles - | - | - | - | 624 |
| Total | - | - | - | - | 3207 |

while there were also irregular forces composed of Burghers, volunteers, half-breeds, Malays, and Hottentots, in all some 10,677 men, making a total of 13,884 officers and men; while in reserve in Uitenhage and Lower Albany were some 3000 Burghers, under Major-General Aylmer.

The drought which prevailed at this time in the eastern districts made the question of transport and supply one of extreme difficulty; but fortunately an anchorage was discovered on the coast a mile east of the Fish River, and only twenty-two miles from the advanced post of Fort Peddie, which enabled this difficulty to be overcome, and early in July the schooner "Waterloo" landed the first cargo of stores at this point, which was thenceforth known as Waterloo Bay.

Subsequently the seamen and marines of H.M.S. "President" arranged a raft communication across the river, protected by a fort, called Fort Dacre, after the rear-admiral commanding the station, thus opening direct communication between Grahamstown and Fort Peddie, protected on its right by the sea, and working round the left flank of the enemy. On the 13th of June Sir Peregrine Maitland established his headquarters at Waterloo Bay. The army of operations, exclusive of the Burghers under Sir Andrew Stockenstrom, was organised in two divisions, the left division under Colonel Hare with headquarters at Block Drift, and the right division under Colonel Somerset, whose headquarters was established at Waterloo Bay.

The Kaffirs had retired with the looted cattle into the fastnesses of the Amatola mountains, and the plan of operations it was intended to pursue was to endeavour to enclose them there and capture the whole. As a preliminary to this object Colonel Somerset moved with his division to re-establish the long-deserted post of Fort Beresford, on the Buffalo River, in the vicinity of the mountains. It was intended that from that position we should block all the outlets from the Amatola mountains to the south and east, while Colonel Hare with the first division and Stockenstrom with his Burghers were to advance on Amatola from the north and west.

This movement, which began at the end of July, proved a failure, as, owing to the impossibility of closing the line, the whole of the Kaffirs got quietly away between Colonel Hare's right and Colonel Somerset's left. Great privations and enormous difficulties were entailed on the troops, who were quite exhausted at the close of the operations, and the only advantage secured was a new advanced position at Fort Cox, twenty miles to the westward of Fort Beresford.

The light company of the 45th, being on the extreme left of the British forces, remained on the Tarka, taking no part in the attempt upon Amatola; but when the failure of the attack in force rendered necessary a system of smaller and separate attacks, it commenced to join in more active service. The first move was made early in August in co-operation with Captain Hogg, of the 7th Dragoon Guards, and was directed against two chiefs whose country lay to the north of Fort Cox, between the Tarka River and Shiloh.

Captain Hogg, who moved from Fort Cox, had with him a detachment of the 91st regiment and 800 Burghers; while Captain Seagram had the 80 men of his own company, a body of Fingoes, some of Kama's Kaffirs, and a small body of Burghers under Commandant Pretorius.

Captain Seagram, leaving his camp on the 19th of August, marched straight upon Mapona's kraal, and after a sharp fight in which nine Kaffirs were killed, took and destroyed it, bringing away about 400 head of cattle. Next day he pushed on to Mapaisa's kraal. The Kaffirs again showed fight, and Captain Seagram, after a sharp contest, defeated them without loss to himself, killing several and putting the rest to flight. On the third day, having fully accomplished his object, he marched to meet Captain Hogg, whom he found in possession of some 3000 head of cattle, which the combined forces brought safely into Shiloh, after a most fatiguing march.

The chief Mapaisa and his tribe, with a quantity of plundered cattle, were supposed to have moved away to the northward after their defeat by Captain Seagram, and some attempts were made by the combined force to get at them, but without success. Captain Seagram consequently returned with his force to Tarka Post.

In the middle of December another opportunity presented itself of attacking Mapaisa, and Lieutenant Garden, with 50 men of the 45th, proceeded again to Shiloh, and was employed under Major Sutton for some time in fatiguing marches into Mapaisa's country, but without any immediate result. Negotiations were soon afterwards entered upon with the smaller chiefs, and Sandile as the paramount chief, as to the results of which disagreements arose between Sir Peregrine Maitland and his subordinates. Sir Andrew Stockenstrom resigned his command in a fury, and Colonel Hare, broken down in health, obtained permission to return home. The year closed a series of transactions, partly political and partly military, which had a superficial appearance of success, but in reality left things very much as they were at the commencement of the war.

The only troubles in Natal at this time were the difficulties caused by the large influx of natives seeking protection from the oppression of surrounding chiefs. Their influx caused a serious effect upon the Dutch population, who trekked away out of the colony in large numbers across the river.

On the 7th January, 1847, a detachment of the regiment, under Captain Blenkinsopp, consisting of 2 subalterns, 1 surgeon, 2 sergeants, and 80 rank and file, accompanied by a detachment of the Royal Artillery with a gun, a few sappers, and 29 men of the Cape Mounted Rifles, marched from Pietermaritzburg to join a force of some 300 men, accompanied by Mr. Theophilus Shepstone, proceeding to the southern boundary of the colony to afford protection to the weaker chiefs, who had applied to the Government to give them aid against the inroads of the powerful chief Foto. The force arrived on the Umkomanzi River on the 11th, and was there detained by floods until materials for the construction of a raft were received from Port Natal.

It was learned on arrival here that Foto had fled, carrying off all his cattle. Though the country was impossible for artillery, and the dense bush had made it exceedingly difficult, yet patrols sent out by the little force succeeded in exacting a fine of 500 head of cattle, and on one occasion met and defeated a party of Kaffirs, killing five and making five prisoners. Foto's kraal was also discovered and destroyed, and the little force returned to Pietermaritzburg.

On the 28th of January, the first anniversary of the battle of Aliwal, the regiment had the honour of entertaining Sir Harry Smith, who had just concluded the pacification of the eastern frontier.

Early in 1848 the light company, which had been encamped at Shiloh since the pacification of the district, rejoined headquarters at Pietermaritzburg, and on their way were the guests of the reserve battalion at Fort Hare.

Before leaving the Tarka Captain Seagram had received

the following very flattering letter from Major Sutton, of the Cape Mounted Rifles, commanding the post:—

"Shiloh, 12th February, 1847.

"Sir,

"In dismissing the force assembled to act under my orders for the present, I cannot allow the 45th detachment to return to quarters without expressing to you my high sense of their extreme good conduct and valuable services during the time they have been under my command. The cheerful, willing way in which they have performed their duty, under circumstances of great privation, hardship, and fatigue, I cannot sufficiently praise. I sincerely regret that the nature of war in which we are engaged did not afford us occasion for them to distinguish themselves in a way which I know would have been much to their satisfaction. Though I have never seen anything better than the service in this country, yet of this service I have had long experience. If, therefore, you think my opinion of sufficient worth to communicate to your men, or that so doing would afford them any gratification, may I beg of you to do me the favour of informing them of it, as also their officer, Lieutenant Garden? At the same time, may I beg of you to accept my best thanks for the ready and efficient way in which you sent the men to the field, as also for your great kindness in sending me the reinforcement, when I knew you could ill spare so many men?—I have, etc.,

(Signed) "WM. SUTTON,

"Major, C. M. Rifles.

"To Captain Seagram, Tarka."

A pleasant little episode marked the march of the company to King Williamstown. It met Sir Harry Smith riding the opposite way, and his Excellency took the opportunity of congratulating the little force on its excellent reputation, recalling his recollection of the regiment in the Peninsula. He had finished his speech and was about riding on when Captain Seagram remembered that one of his men was a prisoner for disobedience of orders. He begged the commander-in-chief to take the opportunity of remitting his punishment, which the general kindly did in a second address, qualifying his pardon with the proviso, " if his comrades wished it." The men all shouted, " We do," and the second parting accordingly took place with mutual feelings of the greatest goodwill. Part of the company embarked at the Buffalo River on the 16th of January, 1848, but the rest, waiting to be joined by some recruits from the reserve battalion, did not leave till the 28th.

In the following September Captain Seagram's services were recognised by his appointment to a brevet majority.

A portion of the regiment in Natal had been mounted. This " mounted troop," as it was called, was first commanded by Lieutenant and Adjutant Gordon, and subsequently by Captain Parish, until its disbandment, after two years of embodiment.

In February, 1848, Major Cooper was detached to form a permanent post on Bushman's River, with the following forces under his command :—140 men of the 45th (including the mounted troop under Captain Parish), a detachment of the Royal Artillery with two guns, and a detachment of the Cape Mounted Rifles. In January, 1849, Mr. West, the Governor of Natal, died, and Lieut.-Colonel Boys was appointed to administer the government, and about the same time Major Cooper was promoted lieut.-colonel to command the reserve battalion quartered at Fort Hare.

Life in Natal now became quiet, and the monotony was only broken by frequent hunting expeditions, which in those days provided good bags; and it is recorded of one party, consisting of Captain Gordon, Lieutenants Griffin and Morris, with Lieutenant Faraday, of the Royal Artillery, that they killed in twenty-two days no less than 137 elephants, 42 buffaloes, 39 elands, 17 rhinoceros, 1 lion, 8 koodoo, and 1 hippopotamus, besides wild boars and other smaller deer.

In August of this year the two battalions of the regiment were again consolidated into one, the 1st battalion becoming the right wing and the reserve battalion the left wing.

# CHAPTER XIII

WE must now return for a while to follow the fortunes of the reserve battalion, which remained at Cork, under the command of Major Butler, after the departure of the 1st battalion. On the 22nd of December, 1843, the reserve battalion received orders to prepare for embarkation for Gibraltar, and the detached companies were accordingly called in; while on the first of the new year the depôt company left, under command of Captain Henry Cooper, to join the depôt battalion at Parkhurst. The battalion sailed, 607 of all ranks, on the "Apollo" on the 12th of January, and arrived at Gibraltar on the 19th. The battalion remained on the Rock for eighteen months, during which time it is recorded that Drummer Brown carried off a prize for skill in drumming in competition with all the rest of the garrison, at the end of which period, on the 31st of July, 1845, it sailed for the Cape of Good Hope on board H.M. troopship "Resistance."

Previous to the embarkation, Sir Robert Wilson issued an order of the day, complimenting the regiment on its general steadiness and soldierlike bearing while in garrison, and also for the assistance it had rendered in the maintenance and repair of the fortifications.

After five weeks at sea the "Resistance" put into Rio de Janeiro for water on the 8th of September, and on the following day the regiment suffered a severe loss in the death of Major Butler, who had commanded the reserve battalion since its formation. He was buried on shore on the 10th, the funeral being attended by a large number of officers from the foreign ships of war lying in the harbour. In two or three days the "Resistance" was again ready to proceed, but the disturbed state of affairs in the River Plate, where British life and property was becoming seriously compromised, caused Mr. Hamilton, H.M. Minister in Brazil, to take upon himself the responsibility of changing her destination, and the regiment was accordingly ordered to Monte Video, where it arrived on the 25th of September, at the same place where it had distinguished itself in the disastrous expedition thirty years before.

The regiment remained on board the "Resistance" until the 21st of October, losing Lieutenant Oakley, who had been fourteen years in the regiment, on the 14th, when it was landed to protect the British subjects in the city, which was closely besieged by General Oribe. The duties which the battalion was called upon to perform were, in company with men landed from the French fleet, and a motley force raised by the Monte Videans from all nationalities, among whom was Giuseppe Garibaldi, were those of a special police, to keep order among the semi-civilised civil population, and had no connection whatever with the defence of the city.

The accommodation afforded them was wretched in the extreme, consisting of old slaughter-houses and sheds out of repair, very damp and infested with rats. At first there was no bedding, and the men had only their blankets and greatcoats; but after a time the quarters were somewhat improved and a certain amount of comfort secured. On the 16th of October, H.M.S. "Apollo" arrived at Monte Video with reinforcements, consisting of the 73rd regiment and a draft of 48 men for the 45th. Their arrival was most opportune, as just at this time a mutiny broke out among the native troops, who might have plundered the city had it not been for the presence of the British troops.

The "Resistance" had meanwhile arrived in England, bringing news of the detention of the regiment at the River Plate. Orders were immediately sent for the corps to proceed to its original destination, the Cape; and accordingly the "Resistance," having brought out Major Cooper to command the battalion, embarked it on the 3rd of July, 1846, after a stay of ten months in South America. The climate and peculiar conditions had told upon the regiment during their stay—two officers and fifteen men had died, and at the time of embarkation over one hundred men were sick with dysentery, ophthalmia, and scrofula.

The regiment had, as usual, maintained its credit under the trying conditions to which it had been exposed; the Government of Banda Oriental sent a special despatch to Mr. Ormsby, the British Minister, bearing testimony to the excellent conduct of the corps; and Mr. Ormsby wrote, in forwarding the despatch, that "the good behaviour of the regiment while serving with the marines and seamen, with French detachments, and with the half-disciplined troops of the Monte Videan Government, was worthy of the highest praise." The Duke of Wellington expressed to the regiment "the great satisfaction" he had derived from the perusal of these testimonies.

The regiment arrived at Simon's Bay on the 30th of July, and at once landed 84 sick, together with all the women and children. On the 8th of August the battalion again put to sea, and disembarked at Port Elizabeth, Algoa Bay, on the 15th of the same month.

The disembarkation strength was 509 rank and file and the following officers:—Major Cooper, Captains Moultrie, Vialls, Tench, Bates, and Parish; Lieutenants Bewes,

M'Crea, Johnstone, and Leach; Ensigns Harvey, Goff, Dawson, and Woodford; and Assistant-Surgeon Barker.

On the 1st of April, owing to a change in the constitution of the reserve battalion, Lieut.-Colonel Archibald Erskine was appointed to the command, and the staff made permanent, with the exception of the paymaster, whose duties were carried out by a subaltern.

On disembarkation the battalion marched to Grahamstown. On the way up Captain Tench accompanied a Mr. de Wet on a shooting excursion; in the bush the party met some hostile Kaffirs driving off some cattle belonging to Mr. de Wet; a brush ensued in which two Kaffirs were killed, and thirteen head of cattle and two horses were recovered.

Grahamstown was reached on the 3rd of September; two companies were ordered to remain there, and on the following day the headquarters marched to join Colonel Somerset's division on the Fish River. They reached Fort Dacre on the 8th, and crossed the river by the ferry on the 10th, joining Sir Peregrine Maitland's camp at Waterloo Bay on the same day. On the 16th of September a patrol of 1500 men, including three companies of the 45th, under Captains Moultrie, Bates, and Parish, operated under Colonel Somerset towards Keiskama Hook and the Buffalo River. The patrol returned on the 3rd of October, having accomplished little beyond the capture of 3000 head of cattle; and having suffered considerable privations from the atrocious weather and the shortness of rations.

Considerable sickness prevailed at this period among the troops east of the Fish River, hardly an officer escaping illness, while provisions were scarce and extremely dear, as much as eighteenpence being paid for a small loaf of bread. In the middle of September H.M. troopship "Apollo" arrived at the Cape from Monte Video, bringing the 73rd regiment and 58 men of the 45th, who had been left behind at that place.   The ship came on to Waterloo Bay with the troops, but was nearly wrecked there, losing three anchors and cables in a heavy gale, and escaping to sea with great difficulty.   The troops in consequence were not landed there, and the detachment of the 45th did not reach camp at Waterloo Bay until the 29th of November.

About the 15th of October Lieut.-Colonel Archibald Erskine joined, and took over command of the battalion; he had probably been on leave in England, as he arrived at the Cape in the transport "Cornwall" from Cork, with Ensigns Fleming and Grantham and a draft of 41 rank and file.

On the 27th of October the headquarters of the battalion quitted the camp at Waterloo Bay, crossed the Fish River and marched to Fort Beaufort *en route* for Fort Hare at Block Drift, which became for some time the post of the battalion; a company of sick and invalids, left behind under Captain Parish, followed the battalion on the 29th of November.   On the 3rd of November a truce of fourteen days was granted to Sandile, and he was offered peace on surrendering 20,000 head of cattle and all his arms.   This he agreed to, and at the expiration of the truce 300 horses and 300 head of cattle were surrendered, as well as the immediate cause of the war, the axe-stealer, and the murderer of the Hottentot guard.

Arrangements were made for registering every Kaffir as a British subject who gave up a musket.   These persons were allowed to settle down in the new British territory; but the new arrangement became somewhat of a farce. Only the worst of the muskets were given up; and for one registered native who settled down, eight or ten unregistered Kaffirs became his companions.   The tribes near the sea were slower in offering themselves than those more inland. Somerset scoured that country in November and brought in some cattle; and by the beginning of December only the chiefs Pato, Kobe, and Toyise were left west of the Kei in open warfare with the colony; but they were subsequently joined by many stragglers from the nominally pacified tribes. The general pacification, however, was supposed to progress, and late in the year the camps east of the Fish River, near the sea, were broken up and the troops were transferred further inland and to the northward. On the 9th of December the detachment which had come out in the "Apollo" followed up the headquarters to Fort Hare. The situation of the camp occupied by the battalion was exceedingly beautiful; it lay on a plain enclosed by bush-covered hills, behind which rose the cloud-capped peaks of the Amatola fastnesses. The river Tyumie wound by the foot of the hills and lighted up the landscape. At the edge of the bush the regiment was camped, and immediately above the camp lay the site of the large fort upon which the battalion was once more exhibiting its traditional constructive genius. This fort, when completed, was intended to accommodate one battalion, one troop of cavalry, one troop of the Cape Mounted Rifles, and a half-battery of artillery.

On the opposite side of the river, at a spot called Lovedale, under the foot of the Amatola range, was a large camp containing the headquarters of the commander-in-chief, and garrisoned by the 27th and 90th regiments, the 7th Dragoon Guards, some Royal Artillery, Cape Mounted Rifles, Fingoes, and Major Hogg's Hottentots.

The men of the regiment, in addition to their work on Fort Hare, which was under the superintendence of Captain Owen, of the Royal Engineers, laid out extensive gardens by the side of the river.

In February a reorganisation of the troops placed the battalion in the "Beaufort Division," under Lieut.-Colonel Johnstone, of the 27th regiment, which also included the 27th and 91st regiments, the 7th Dragoon Guards, and a Burgher force under Major Sutton of the Cape Mounted Rifles.

On the 7th of January Sir Peregrine Maitland, who had been superseded by Sir Henry Pottinger, had started for Cape Town, leaving Colonel Somerset in command; but before leaving, believing the war to be over, he had abolished martial law in the district and sent the 90th regiment to Port Elizabeth to embark for England.

Sir Henry Pottinger, who had appointed Captain Fellowes, of the 45th regiment, his A.D.C., arrived at Grahamstown on the 28th of February, and, learning the condition of things, at once countermanded the orders to the 90th regiment, took steps to raise new levies, and determined to push the frontier to the line of the Buffalo River, along which a chain of posts was ordered to be established, and its mouth was selected as a landing place for supplies instead of Waterloo Bay.

Early in February the Kaffirs carried off some cattle from the camp at Block Drift, and a patrol, which included a party of the 45th, which went in pursuit, succeeded in bringing in a quantity of cattle as a compensatory fine. Sir George Berkeley, who was now appointed to command the troops, took a more serious view of the general situation, and, accordingly, proceeded to increase the forces to the utmost of his ability. The Cape Mounted Rifles were increased, Captain Hogg, of the 7th Dragoon Guards, was employed to raise a force of Hottentots, and orders were sent to Cape Town to detain troops on their way home from India.

Such was the state of affairs when, early in June, fourteen goats were stolen from the Kat River and traced to one of Sandile's kraals. Mr. Calderwood, the Gaika commissioner, thereupon required from the chief the restitution of the stolen property, a fine of three head of cattle, and the surrender of the thief. Sandile confiscated the property of the thief and his friends, but only complied with the demand to the extent of restoring the goats. Sir H. Pottinger, upon this, determined to arrest Sandile; for this purpose Lieutenant Davies was despatched on the 18th of June, with two officers and 74 men of the Kaffir police, assisted by 100 men of the 45th, 50 men of the 7th Dragoon Guards, a small party of the Cape Mounted Rifles, and some Fingoes, under the command of Captain Moultrie, of the 45th.

On reaching Sandile's kraal, near Burnside, it was found that he had fled, so two horses and about thirty-nine head of cattle were seized; in a short time, however, very large bodies of Kaffirs were seen advancing, Sandile amongst them, and the small force deemed it prudent to fall back. The Kaffirs followed up nearly to Block Drift, keeping up a running fight all the way. One private of the 45th, named Grimes, was killed, as well as another member of the force, and four were wounded. In a few days Sandile repented and sent in twenty-one head of cattle as a peace offering; but the Governor decided that the chief must surrender. This he declined to do, so preparations were made for a renewal of hostilities. On the 18th of August, everything being in readiness, a police escort was sent to Sandile to demand the surrender of the thief who stole the goats, and two hundred muskets. As was expected, Sandile treated the messenger with disdain; so on the 27th of August Sir Henry Pottinger proclaimed him a rebel, and called out the Burghers to aid in attacking him, offering them all the cattle they could seize in his district. Very few Burghers, however, turned out, as they regarded an attack on Sandile, under the circumstances, as perfectly useless.

Nevertheless, Sir George Berkeley determined to attack Sandile with all the irregulars he could muster, and all the regulars that could be safely withdrawn from the posts of defence. Fort Hare being completed, the troops were

M

drawn across the river and accommodated in the huts within its enclosure. This was made the first depôt, there being a good road from Waterloo Bay, about fifty miles distant. Fort White, which was of a very similar character, twelve miles to the westward, was made the second depôt; and here Sir George Berkeley took up his headquarters. The third depôt was at King Williamstown, fifteen miles east-south-east of Fort White. North of the Amatola stronghold a depôt was formed at Shiloh, where a large irregular force under Captain Sutton was stationed, to prevent the Kaffirs escaping in that direction. The Kaffirs met these preparations most effectively; they quietly drove their cattle and people down into the settled districts, mixing them up with the tribes that had submitted, thus leaving the troops nothing to get hold of in Amatola, and enabling the chiefs to keep in hiding.

On the 19th of September the first patrols were sent out from the three southern posts, in light marching order. They numbered in all about two thousand men, and were commanded by Colonel Somerset, Lieut.-Colonel Buller, of the Rifle Brigade, and Lieut.-Colonel Campbell, of the reserve battalion of the 91st regiment. The 45th was broken up into detachments. One party of 170 men, under Major Hind, marched from Fort Hare as part of Colonel Campbell's force; while another detachment proceeded to Post Victoria to protect the colonial boundary. Part of Colonel Campbell's column, including the 45th detachment, was almost the only portion of the force that had an encounter with the enemy. This took place at Salamba's Kop, near Fort White. The column had marched at four o'clock in the morning, Lieutenant Dawson, of the 45th, leading the van. A large party of Kaffirs were discovered squatting round an iron pot; hoping to surprise them the column diverged into the bush and marched in silence. A musket, however, accidentally fired gave the alarm, and the natives fled, leaving their iron pot behind them, which was seized by a man of the regiment named Harris, who carried it on his back. The Kaffirs, who had concealed themselves in the bush, soon began to open fire on the column, and a bullet destined for the man who had taken possession of the pot was stopped by the pot itself, which saved his life. A song descriptive of the event was long popular in the regiment.

Although little was effected by force of arms, the chiefs were soon weary of the struggle, and in less than a month Sandile, on condition that his life was spared, on the 19th of October surrendered to Colonel Buller, near Keiskama Hoek. This completed the pacification of the Amatola district, and Sir George Berkeley moved his troops up the Kei River against Pato and Kreli. These chiefs soon surrendered, and by the 19th of December the "war of the axe" was at an end.

Meanwhile Sir Henry Pottinger and Sir George Berkeley had been appointed respectively governor and commander-in-chief in Madras; and Sir Henry Smith, the victor of Aliwal, arrived at the Cape as governor and commander-in-chief.

The new governor lost no time in proceeding to the frontier. He formally ordered all hostilities to cease, and assembling the chiefs he made arrangements with them by which he hoped to secure a permanent pacification along the whole frontier.

The beginning of 1848 found the reserve battalion concentrated at Fort Hare, with one company at Fort Cox, and a small detachment at Fort White. On the 23rd of February a working party of 42 rank and file and a sergeant, under command of a subaltern, were sent to repair the middle drift of the Keiskama River, and remained so employed until the middle of August.

On the 31st of March Colonel George MacKinnon, commanding the troops in British Kaffraria, inspected the battalion at Fort Hare, expressing the usual favourable opinion as to its condition.

On the 18th of July the battalion lost its commanding officer, Lieut.-Colonel A. Erskine, from disease of the brain, at the early age of forty-one; he was succeeded by Major Cooper, then in Natal with the 1st battalion.

Trouble, which had been brewing for some time, now began to arise across the Orange River; difficulties between the native tribes themselves, and between them and the immigrating Dutch farmers, pointed steadily to the early interference of the Colonial authorities. Ever since 1845 British troops had been on the Orange River endeavouring to keep the peace, and collisions had occurred between them and the farmers.

At the end of 1846 Mr. Pretorius, as the delegate of the Dutch farmers who had immigrated from Natal, sought and was refused an interview with Sir Henry Pottinger, to represent the grievances of his countrymen; in despair he led the way in a fresh emigration of farmers, hostile to the Queen's Government, across the Orange River.

After arranging affairs on the eastern frontier, Sir Harry Smith proceeded to undertake the settlement of matters across the Orange River. In January, 1848, he held a meeting at Bloemfontein with the chiefs of the contending native tribes, and many heads of the loyal Dutch families. His intention was to substitute the Queen's government for the anarchical system prevailing, but he does not seem to have recognised that to do this required military force. From Bloemfontein he proceeded to Winburg, where he met the powerful Basuto chief Moshesh; here, also, many of the loyal Dutch waited on him and pleaded for the establishment of the Queen's government. While there he received news

Map of the
ORANGE RIVER
SOVEREIGNTY.

Kathlamba Mountains or Drakensbergen

Wilge River

Vaal River

Valse River

Sand River

Vet River

Modder River

Kaal Spruit

Riet River

Orange River

Caledon

Buffels Vlei

European Villages.
French Mission Stations.
Wesleyan.
Berlin.
London.    Station.

of the approach of Pretorius and the emigrants from Natal;
he at once rode off to meet them, and was received by them
with all respect.

Pretorius, however, firmly declined to return to Natal or
accept Sir Henry's plan of establishing the British government over the district; and as a result of the governor's
proclamation of the 3rd of February, 1848, declaring the
sovereignty of Her Majesty from the Orange River to the
Vaal River, and eastward to the Kathlamba Mountains, a
large number of the Dutch farmers moved northward across
the Vaal River, while Pretorius set himself to agitate and
prepare for overt resistance.

In March Mr. Biddulph was appointed civil commissioner
of the new district, which was called "The Orange River
Sovereignty," with his headquarters at Winburg; but the
farmers of the district at once threatened revolt.  Sir Henry
issued a proclamation against them, to which they replied
by manifestoes declaring their desire for independence,
though a minority declared their allegiance to the Queen.
The Burghers beyond the Vaal were much in sympathy with
the malcontents of Winburg.  They held a meeting at
Potchefstroom and passed threatening resolutions against the
governor's proclamation.

When Mr. Biddulph was installed as resident by Major
Warden at Winburg, the disloyal inhabitants of the district
declined to obey him, and sent messages to Pretorius, declaring their intention of fighting for their independence, and
demanding his assistance.

Mr. Biddulph, under threats of arrest by the Burghers,
shortly afterwards retired from Winburg to Bloemfontein;
and on the 12th of July Pretorius, at the head of an armed
force, arrived at Winburg; here he declared that all farmers
who would not join him must cross the Orange River.  Some
of the loyal farmers defied him and threw themselves into

laager, but the greater part joined him.   He at once confiscated the property of all the loyal men and sent out
patrolling parties to overawe the district, one of which nearly
captured Major Warden and Mr. Biddulph about six miles
from Bloemfontein.

On the 17th of July Pretorius, at the head of more than a
thousand men, arrived within two miles of Bloemfontein,
and Major Warden, who had but fifty-seven men of the Cape
Mounted Rifles under his command, retired upon Colesburg.
Pretorius took formal possession of the town on the 20th
of July, and then advanced and formed a camp on the north
side of the Orange River.

Major Warden's report reached Sir Harry Smith at Cape
Town on the 22nd of July.   He at once issued a proclamation
offering a reward of £1000 for the apprehension of Pretorius,
and issued orders for all available troops to assemble at once
at Colesburg.   He also started for the frontier immediately,
and arrived at Colesburg on the 9th of August.

Among the troops ordered to Colesburg were two companies
of the reserve battalion, numbering 170 men, under command
of Captain Blenkinsopp, with Captain Tench, Lieutenant
Dawson, and Ensigns Fleming and Howard.   The detachment marched from Fort Hare on the 29th of July and
reached its destination on the 21st of August.

The whole force assembled under the command of Sir
Harry Smith consisted of about 800 men, made up of the
above detachment, four companies of the Cape Mounted
Rifles, two companies of the Rifle Brigade, two companies
of the 91st regiment, a few engineers, and some artillery
with three six-pounders, and the force was subsequently
joined by a few loyal Dutch farmers and about 250 mounted
Griquas.

On the 22nd of August the passage of the Orange River
commenced, and though it took five days to complete, only

a single raft being available, it was not interfered with by the enemy, who had begun to fall back upon Bloemfontein. On the 27th the force moved forward to Philippolis, the infantry being under the command of Major Beckwith, the artillery Lieutenant Dynely, and the whole under Lieut.-Colonel Buller, of the Rifle Brigade, under the immediate orders of the commander-in-chief.

The next day the march was continued to Visser's Hoek, the country through which the force marched having been completely abandoned by its inhabitants. That evening a party of farmers who went ahead as scouts returned with the information that the country was clear as far as Boomplaats, some fifteen miles in advance. The column moved forward again at dawn the next morning, halting for their morning meal at a place called Touwfontein. The Cape corps was pushed on as an advanced guard, supported by the two companies of the Rifle Brigade, behind whom marched the engineers and the artillery with their guns, followed by the 45th and the 91st detachments; while the rear was brought up by a long train of waggons laden with stores and ammunition. In this order the column traversed the open plain which stretches to within a few hundred yards of the Kromme-Elleboog River.

Here the features of the country changed. A chain of hills, strewn with boulders, rose on the right side of the road, and beyond the river rose another chain of hills in a valley below which lay the farmhouse of Boomplaats; while beyond lay a third chain of hills cut by a neck through which the road passed to the plain beyond.

Information was received from a shepherd that the Boer forces had passed the night at Boomplaats, and Lieutenant Warren, with a couple of men, was pushed ahead to reconnoitre. He returned in a few minutes, reporting that he had seen the farmers in considerable force beyond the nearest range of hills. Lieutenant Salis, with a troop of the Cape corps, was then ordered to ride on ahead of the main column, and Sir Harry Smith himself galloped out to the front eager for a parley with the emigrant leaders in the hope of avoiding a collision. It was about 11 a.m. when the party came abreast of the second hill on their right, when someone exclaimed, "There they are." In an instant the crest was covered with men, and a volley, followed immediately by a second, greeted the party, which instantly turned and galloped back to the main body. The governor's horse was wounded in the face, and one of his stirrup leathers cut through; Lieutenant Salis was badly wounded in two places, his horse shot dead, and three Hottentots killed.

The governor at once gave orders for the guns to move up into position, and under cover of their fire the Rifle Brigade and 45th detachments carried the position by assault under a storm of bullets. Driven from this position the farmers fell back towards the centre of their line and prepared to make another stand at the next hill.

In the meantime a party under the command of Commandant John Kock made a determined attack upon the waggons and supplies, but the defence offered by the Cape corps and the loyal farmers was sufficient to compel Kock to retire, in doing which some loss was inflicted upon his force by the guns, whose fire he had to cross.

The artillery now moved forward to a new position and the whole force advanced, the companies of the 91st regiment being sent forward to reinforce the Rifle Brigade and the 45th. Each position in turn was carried, the only obstinate resistance being encountered at a stone cattle kraal at Boomplaat's farm. Driven from this, the farmers made a final stand on the slopes commanding the neck traversed by the road. The first attack by the Cape corps and mounted Griquas failed, but the infantry being brought up, the position was stormed, and the enemy fled over the plain to the eastward. Such was the affair of Boomplaats, which Sir Harry Smith described as "one of the most serious skirmishes that had ever, he believed, been witnessed."

The two companies of the 45th were in the front of it from first to last, and lost three men killed and nineteen wounded, including Captains Blenkinsopp and Tench. The total loss on the British side was two officers killed—Captain Murray, of the Rifle Brigade, and Ensign Steele, of the Cape Mounted Rifles—and forty-nine wounded, among whom was Lieut.-Colonel Buller.

Among the killed was Bugler Baylis, of the 45th, who was shot while sounding the charge at the commencement of the action. "His officer, Lieutenant Fleming, took the bugle from his body, and, slinging it on his own shoulder, thus guided his company."

The farmhouse of Boomplaats was converted into a hospital, and the next day the troops pushed on to Bethany, a station on the Riet River. Various accounts are given of the losses of the rebel farmers, but the one generally accepted is nine killed and five wounded.

The next day the force pushed on to Bethany, a station founded by the Berlin Missionary Society in 1835. On the march the Griqua scouts captured two stragglers, who had taken part in the fight at Boomplaats. One was a deserter from the 45th named Quigley; he had sent notice of the Boer movements to the British resident, but this second piece of treachery did not save him. He was tried by court-martial at Bloemfontein and shot. The other prisoner was a young Dutch farmer named Dreyer, who was, by what appears to have been a grave error of judgment, also tried and shot.

At Bethany Sir Harry Smith issued a general order to the troops in which he congratulated them "upon the brilliant achievement in dislodging from one of the strongest positions ever attacked the rebel force, consisting of nearly a thousand men, well armed and organised, though in a bad cause. The result has been a complete dispersion of the rebels." It was always the custom to shoot the bullocks required for the men's rations, and on the march from Boomplaats Private Hunt, of the 45th, was detached for this duty. On one occasion he missed the animal's head, and the bullet glancing off one of its horns, killed a Griqua who was standing near. He was made prisoner, and taken before Sir Harry Smith, and it having been shown that he was the first man to crown the hill at Boomplaats no more was said, and the Griqua chief was compensated and consequently pacified.

The force rested at Bethany for two days, having marched forty-six miles on the day preceding their arrival there, and then resumed their march to Bloemfontein, which was reached on the 2nd of September. Here Sir Harry Smith issued proclamations confiscating the property of the insurgents, and offering rewards for the apprehension of the ringleaders, £2000 being offered for Commandant-General Pretorius. From Bloemfontein the troops marched to Winburg, which was occupied without any further opposition on the 7th of September. Here Sir Harry Smith, under a salute of twenty-one guns, proclaimed the Queen's authority from the Orange to the Vaal Rivers, and divided the territory into four districts, Bloemfontein, Caledon River, Winburg, and Vaal River.

The troops soon retraced their steps to Bloemfontein, most of the force with Sir Harry Smith continuing the march across the Orange Rver. The two companies of the 45th, with one company of the Cape Mounted Rifles, and 25 men of the Royal Artillery, with three six-pounders were, however, left behind to garrison Bloemfontein, under the command of Captain Blenkinsopp, who was soon afterwards rewarded with a brevet-majority for his services. On the

15th of September Sir Harry Smith published at Bloemfontein a farewell order to the troops, in which he declared
he had "the greatest satisfaction and happiness in again
recording his high sense of their gallantry as soldiers, of the
patient endurance of fatigue in the long-continued marches
they had made, and of their exertions in crossing rivers, and
other laborious duties which attach to military operations."
"His Excellency," said the order, "has served with many
troops, but in no campaign has he been associated with more
energetic officers and soldiers than those composing these
detachments."

He ordered that the troops who had been employed on Fort
Victoria, chiefly the 45th regiment, should receive working
pay at the rate of fourpence a day, a very unusual grant in
those days; troops employed on such works, as a rule,
received nothing extra.

In conversation with the men of the regiment Sir Harry
Smith found that two of them had forfeited their service for
misconduct; he undertook and ultimately succeeded in
obtaining a remission of their penalties as a reward for their
gallantry at Boomplaats.

Sir Harry Smith, on his way to the Cape, visited the
remainder of the reserve battalion.   At Fort Hare he
delivered an address to the wives of the regiment, and
especially to those whose husbands had been wounded at
Boomplaats, assuring them that they would be taken care of,
and praising their gallantry.   At Fort Cox he addressed the
detachment, and, speaking of their comrades who were at
Boomplaats, said, "When you write tell them how proud
their commander-in-chief is of them."

At Fort White he spoke to the detachment as follows:—
"Soldiers of Her Majesty's Fighting Forty-fifth! I am
delighted to see you; you are an honour to Her Majesty's
service and to your country.   Your appearance does you and

your commanding officer the highest credit. You have, in the arduous services in which you have been engaged, done your duty in your old way, and behaved, as you always have done, like British soldiers. More I need not say. My men! I love a soldier! and if you love your general as he does those under his command we cannot but get on well together. At the same time let me caution you not to abuse an indulgence. I am about to order the re-issue of spirits, but take care you don't run into excess. I have been a soldier forty-two years and have drunk my glass of grog, but never got drunk in my life. Behave like men, and show that you can enjoy in moderation the indulgence I am about to grant you."

At the opening of 1849 Captain Moultrie was in temporary command of the battalion, but on his selling-out the command devolved on Major Kyle until Lieut.-Colonel Cooper arrived from the first battalion on the 29th of March and took over permanent command.

Poor Captain Moultrie did not long survive his retirement. On the 20th of August he left Fort Beresford shortly before sunset, on horseback, for Block Drift, and was never seen alive afterwards. His body was found next morning in the river, caught in a tree, below Corpentas Drift. He had evidently been swept away while crossing the stream, his horse being found on the other side with saddle and bridle saturated with water. He had previously served at the Cape in the 75th regiment, and was well known for his devotion to sport and his knowledge of the country.

There is nothing more to chronicle in the history of the reserve battalion, which, as we have already seen, was in 1850 merged once more into the first battalion, then serving in Natal.

# CHAPTER XIV

THE pacification of the white population of the Orange River
Sovereignty having been thus accomplished by force of arms,
the attention of the Government began to be turned towards
the state of affairs existing among the native tribes, which
was assuming a serious aspect. The south-eastern portion
of the sovereignty, lying chiefly between the Orange and
Caledon Rivers, had been set apart for the occupation of the
three powerful and rival chiefs, Moshesh, Sikonyala, and
Moroko, of which the former was by far the most powerful.
Between them now arose feuds over boundaries, and, as a
consequence, cattle-lifting and murders were of daily
occurrence. The Burghers declined to obey the calls to
arms of Major Warden for the purpose of interference in
these disputes, and the resident had been compelled to adopt
the unsatisfactory policy of endeavouring to effect such
combinations among the blacks themselves as might tend to
overawe the more powerful chiefs, which finally resolved
itself into a struggle between the British resident and the
most powerful of the chiefs, Moshesh.

By the middle of 1851 Major Warden had practically lost
all authority in the sovereignty, and could only take up a
defensive position at Bloemfontein, from which place he
sent to Governor Pine of Natal asking for military aid. In
response to his request a force was sent from Natal under
command of Captain Parish, of the 45th, consisting of two
companies of the regiment under Lieutenants Morris, Miller,

Grantham, and Ensign Barnes; 14 men of the Cape Mounted
Rifles under Lieutenant Mill, and 700 Zulus under Mr.
Ringler Thomson. Deputy-Commissary General M'Clintock
and Assistant-Surgeon Sparrow accompanied the force,
numbering in all some nine hundred men.

Captain Parish was directed to march *via* Ladysmith to
Harrismith, where it was expected he would meet guides and
receive further instructions. But in any case he was directed
to open up communication with Major Warden at the
earliest opportunity, and at the same time to keep Colonel
Boys informed of all his proceedings. Detailed arrangements
for his march were furnished to him; the ammunition
waggon was to be placed nearest the column, and the rest of
the waggon train in rear of it; the march was not to be
pressed so as to impair the efficiency of the cattle, and, above
all, the difficulty of obtaining supplies was not to be lost
sight of. After a difficult march the column reached Bushman's River on the 8th of August and encamped at Nelson's
Kop. During the march the number of friendly Zulus had
increased, and the force now consisted in all of about 1100
with nineteen waggons.

On the 23rd of August Captain Parish received a letter
from Mr. Biddulph at Winburg warning him of a new and
unexpected danger. It informed him that the Boers along
his line of route were on the verge of an outbreak, and that
he must take especial precautions against being misled by
reports specially framed to lead him into difficulties. It was
held at that time that, but for the able precautions taken
by Captain Parish in consequence of this information, the
disaffected farmers would have connived at the native
schemes for cutting off the little column.

Captain Parish arrived safely at Winburg on the 31st of
August, where he received letters from Major Warden
informing him that owing to the disaffection existing among

the farmers no operations could be undertaken against the
natives until the arrival of reinforcements.  He therefore
remained at Winburg with the European portion of his force,
while Mr. Thornton, with the native contingent, continued
the march to Bloemfontein.

Captain Parish was in a somewhat critical position at
Winburg.  It is true he received, on the 20th of September,
a loyal address from the farmers, but the document only
bore twenty signatures, and it was understood that no
practical assistance would be forthcoming in the case of any
coalition between the disaffected Boers and the natives.
Cattle and horses were constantly being carried off, and the
postal arrangements were most irregular, and often failed
altogether.  He had made his camp as defensible as possible
by means of stone walls, but he went through an anxious
time until receiving, on the 1st of October, orders to proceed
to Bloemfontein.

After a weary march of four days through a dead level of
parched desert, under a burning sun, the little force reached
Bloemfontein, and was played in by a drum and a fife belonging to the one company of the left wing of the regiment,
which had so long been detached there under Captain Bates.

The two companies resided under canvas at Bloemfontein,
which was then but a mere village, and Captain Parish had
the pleasure of receiving an order of approval from Sir Harry
Smith of his proceedings as far as Winburg, in which " the
able and energetic manner in which the march had been
conducted " was mentioned.

On the 18th of November Captain Parish, with his force,
accompanied by two six-pounders, marched from Bloemfontein to Thaba Nchu, the residence of the chief Moroko,
who was a supposed ally, and whose protection against
Moshesh was part of the resident's policy.

Important movements among the Boer population now

began to determine the situation.   The hostilities which were
in progress on the other side of the Orange River, and which
were taxing the powers of the Government to the utmost,
were held by the disloyal Boers to present a favourable opportunity to assert their independence.   On the 25th of August
they sent a request from Winburg to the proscribed Pretorius
to assume the office of Administrator-General.   By way of
answer Pretorius sent a letter on the 9th of September to
Major Warden announcing that, at the request of Moshesh
and other chiefs, as well as of many of the white inhabitants,
he had undertaken to proceed to the sovereignty to devise
measures for the restoration of peace.

Through want of military force Major Warden was then
placed in a most humiliating position.   He had no means
of asserting the will of the Government, and was obliged, in
consequence, to send two assistant commissioners to treat
with Pretorius.   In consequence, Captain Parish was ordered
to return to Winburg to keep order and protect the loyal
inhabitants.

The meeting ultimately took place on the Sand River on
the 16th of January, 1852.   Pretorius was attended by three
hundred Transvaal Boers and one hundred of the disloyal
farmers of the sovereignty.   The two assistant commissioners, Major Hogg and Mr. Mostyn Owen, were escorted by
one hundred men of the 45th under Captain Parish, supported
by thirty-five of the Cape Mounted Rifles and one field piece.
As a result, what is called the Sand River Convention was
signed, which guaranteed the independence of the Transvaal
Boers.

Before the close of 1851 a reinforcement of another party
of the 45th, under Lieutenant Coxon, and a few of the Cape
Mounted Rifles, had joined Captain Parish at Winburg, but
were ordered back again to Natal, the party of the 45th
leaving Winburg on the 28th of December.   At Lieutenant

Coxon's first camping ground on the Sand River he was surprised by a party of bushmen, who carried off 82 of his draught cattle.   On hearing the news Captain Parish at once despatched a party of the Cape Mounted Rifles to assist him. On the 30th of December Lieutenant Coxon, with eighteen men, followed up the trail of the cattle, but the bushmen were successful in defending themselves, and Lieutenant Coxon was severely wounded in the encounter which ensued. Captain Parish then sent a further reinforcement of thirty men, who escorted the party back to Winburg.

The disloyalty of the farmers about Winburg now became more and more openly expressed, and on the 7th of February a detachment of the 45th, with a few Cape Mounted Riflemen, was sent to make a raid upon a farm in the hopes of capturing some of the conspirators.   The place, on arrival, was found deserted, and the expedition returned with but one prisoner, who was taken, with his waggons, *en route*.

Early in March, 1852, Sir Harry Smith was succeeded in the government of the Cape by Lieut.-General the Hon. George Cathcart, and Mr. Darling was appointed Lieut.-Governor of the Cape Colony.   On the 6th of April the resident at Bloemfontein received from the Lieut.-Governor of Natal a request for the return of Captain Parish's force, but it was not until the 5th of June that it quitted Winburg and rejoined headquarters at Pietermaritzburg on the 26th. Captain Parish shortly after sailed for England, having been detailed for duty with the depôt at Chatham.

Sir George Cathcart, who had marched a considerable force across the Orange River for the subjugation of Moshesh, was practically defeated by him on the slopes of the Berea mountain on the 20th of December.   As a result the abandonment of the Orange River Sovereignty was determined upon, and on the 30th of January, 1854, a proclamation was signed renouncing all dominion and

N

sovereignty over the Orange River territory.  On the 11th
of March of the same year the British flag was hoisted for
the last time over the Queen's Fort at Bloemfontein, and
Captain Bates, with his detachment of the 45th, who had
been so long in garrison there, marched for Fort Hare.
While these events had been proceeding beyond the Orange
River another Kaffir war had been carried on on the frontier
of the Cape Colony in British Kaffraria.

In the early autumn of 1850 the colonists of that district
urged upon Sir Harry Smith, who had arrived at the frontier,
the necessity of deposing the whole of the native chiefs and
the substitution of British control.  He was, however, by no
means disposed to admit that the fears of the colonists were
justified, or that there was any necessity for the taking of
such measures as they suggested, considering that the
system he had established, which placed the chiefs in nominal
power, taking away their substantive power, was quite
sufficient to keep them under control and secure peace and
tranquillity.  A brave, energetic, loyal, and enthusiastic man,
he lacked one quality absolutely necessary in dealing with
men, and especially with native tribes—the knowledge of
mankind; and forgot, moreover, that the one thing essential
to the success of his system was force to complete it, an
element which was entirely lacking at that time on the
frontier.  On the 8th of November he left Grahamstown for
Cape Town under the impression that his unexpected visit
to the frontier had had the effect of quieting matters and
allaying the excitement and unrest which prevailed there.
On the 26th of November he wrote from Cape Town that he
had left " British Kaffraria in a state of perfect tranquillity."
Only nine days later he had to write that "the quiet he had
reported in Kaffraria, and which he had so much reason to
anticipate, was not realised," and the same evening he
started in H.M.S. " Hermes " for East London, taking with

him the 73rd regiment and some artillery.  War had broken
out which was to continue with considerable fierceness for
upwards of two years.

On the 9th of December, 1850, he was at King William's
Town with the 73rd regiment, under Lieut.-Colonel Eyre,
and 30 men of the Royal Artillery.  Even then he failed to
realise either what was before him or the utter insufficiency
of his force, but still hoped to overawe the natives, who, he
knew, were accurately informed of all his movements, by a
demonstration against the tribes in the Amatola mountains.
Including garrisons, the total force under Sir H. Smith could
not have exceeded 2400 men, composed nearly as follows :—

| | | | | |
|---|---|---|---|---|
| Royal Artillery | - | - | - | 90 men, with 5 guns. |
| Engineers | - | - | - | 192 |
| 6th Regiment | - | - | - | 560 |
| 45th Regiment (Left Wing) | - | | | 452 |
| 73rd Regiment | - | - | - | 530 |
| 91st Regiment | - | - | - | 521 |
| | | | | ——— |
| Total | - | - | - | 2345 |

Out of this meagre force garrisons for no less than
fourteen posts were required, some of them in the heart of
the enemy's country.   Sir Harry Smith held a meeting
of native chiefs at King William's Town on the 14th of
December, at which Pato and the other chiefs from the open
country near the sea emphatically asserted their loyalty.
Word was sent to Colonel MacKinnon to summon the Gaika
chiefs to meet at Fort Cox on the 19th, and on the 16th the
whole available force marched in two columns from
King William's Town and Grahamstown upon the Amatolas.
No part of the 45th was comprised in this force, the whole
wing of the regiment being distributed at this time in
garrison at Forts Cox, Hare, and White.

All the chiefs assembled on the 19th of December, except Sandile, at Fort Cox, and, as usual, agreed to everything and promised everything.   Sir Harry Smith expressed himself as " satisfied with the prospect," but six days later his opinion changed, and on the 26th of December he wrote—" The state of affairs in British Kaffraria is critical."

This change came about as follows :—A force was despatched under Colonel MacKinnon, consisting of detachments of the 6th and 73rd regiments, with some of the Cape Mounted Rifles, in the direction of Sandile's supposed place of concealment.   All went well until the infantry reached a narrow rocky pass, through which the Cape Mounted Rifles had already safely passed, when from behind every rock and bush a heavy fire was opened upon the long and attenuated column.   Twenty-one men were killed and wounded, and a long and arduous struggle ensued before the bush could be cleared of the enemy.   Colonel MacKinnon now determined to retire to Fort Cox, fearing a spread of open insurrection, and accordingly marched on Christmas Day to the southeastward along a waggon road which led to the open country. Sandile with his forces lined the heights on either side, and kept up a fire upon the column during the greater part of the march, inflicting material loss.   On reaching Debe Nek, about two miles from Fort White, the force was appalled at finding the horribly mutilated bodies of a sergeant and fourteen men of the 45th.   Having left his infantry, who were much exhausted, at Fort White, Colonel MacKinnon pushed on with the Cape Mounted Rifles and Kaffir Police to Fort Cox, where he made his report to the governor.

Meanwhile, on the 22nd of December, three officers, five sergeants, and seventy-nine rank and file of the 45th had marched from Fort Hare to Fort Cox, and it was to this party that the unfortunate men massacred at Debe Nek belonged. A corporal and three men had been sent to escort some sheep back into Fort White; they did not return when expected, and it was reported that they had been killed. Lieutenant Goff, of the 45th, then sent out a party of twelve men with a waggon to bring in their bodies. A Hottentot waggoner who escaped reported to Colonel MacKinnon that a large body of Kaffirs, armed with assegais, had suddenly fallen upon them and massacred the whole party before they had time to form up for defence. Emboldened by this success the Kaffirs had even approached Fort White, carried off some cattle, and faced the fire of the garrison until alarmed by the approach of Colonel MacKinnon's force. Lieutenant Goff, of the 45th, was much commended for his gallantry and conduct during this attack.

Things now began to assume a most serious aspect. On the 28th of December Colonel Somerset forwarded despatches to Sir Harry Smith at Fort Cox, but the messengers returned, the road being closed by masses of Kaffirs, with the news that Sir Harry Smith was cut off and blockaded there. The next day he started with 150 men of the 91st regiment, seventy of the Cape Mounted Rifles, and some artillerymen with a three-pounder to attempt his relief, but was met by an overwhelming force of Kaffirs and driven back with heavy losses, including the gun, which, however, was recovered the next day. His retirement on Fort Hare was covered by 100 men of the 45th, under Lieutenant Griffin, sent out by Major Forbes, of the 91st, who had been left in command. Colonel Somerset's loss during this action was twenty killed and fifteen wounded.

The situation now indeed seemed desperate, and the whole frontier was thrown into a state of alarm and panic. The Kaffir police had gone bodily over to the enemy, all the military villages peopled by discharged soldiers, except Ely, had been destroyed and the inhabitants massacred; while the farmers, instead of mustering for defence, were abandoning their properties and flocking inland for protection.

Happily, things, when at their darkest, began to mend. Colonel Somerset got a message into Fort Cox by a loyal Kaffir, and received a reply which authorised the raising and paying of forces; and on the 9th of January it was known that Sir Harry Smith had escaped and got into King William's Town with very little loss.

On the other hand, the Kaffir chief, Hermanus, had induced the bulk of the Kat River Hottentots to revolt, and under his leadership they were sweeping the Fort Beaufort district. Kreili, the great trans-Kei chief, and Madaesa, chief of the Tambookies to the north, had revolted, adding many thousands to the fighting strength of the Gaikas.

On the 7th of January Fort Beaufort was attacked by the Hottentots under Hermanus, but they were beaten off by the garrison, consisting of the Fort Beaufort volunteers and a force of Fingoes, under Colonel Sutton, and their leader killed. On the 28th of January a force of some two or three thousand Kaffirs made a spirited attack on the cattle round Fort Hare, pressing their attack right up to the guns of the fort, but were repulsed with heavy loss; while on the following day they had the temerity to threaten King William's Town itself, but were again driven off.

By the 25th of January 1650 Hottentot volunteers had joined the governor at King William's Town, but much difficulty was caused by the Dutch farmers, who, in spite of the proclamation of martial law, could not be induced to move towards the frontier. Some difficulty was experienced in arranging for the details of the financing of the new levies called up, but this was got over by the appointment of Mr. Cassidy, a retired quartermaster of the 45th regiment, who was given the pay and allowances of a captain while employed in his somewhat arduous task.

On the 30th of January Colonel MacKinnon, with 2200 men of the 73rd regiment, Cape Mounted Rifles, Hottentot and Fingoe levies, and one gun, marched for Forts White and Cox, and, notwithstanding considerable opposition, returned to King William's Town on the 1st of February without loss. On the 3rd of February he again marched, and raided Segoli's territory, by way of Line Drift, burnt his kraal and that of Kinlangini. Again, between the 13th and 19th February, Colonel MacKinnon marched with a force of 2500 infantry to Fort Hare, by way of Fort White, for the purpose of reinforcing Colonel Somerset. The force was threatened, but not seriously attacked, along the line of march. The Tyumie valley was then patrolled and crops destroyed without opposition; but on the return journey to Fort White the Kaffirs continuously molested the column, inflicting on it a loss of twenty-five killed and wounded. Although the Kaffirs had been defeated with considerable loss in their several attacks on British columns and posts they were still powerful and numerous, and the embittered feeling engendered was kept alive by the preaching of the prophet Nonlangini.

During the progress of these troubles the left wing of the 45th was distributed in garrison in the advanced posts, and seldom took part in the operations of the heavier patrolling columns. In May, 1851, it was distributed as follows, excluding invalids at Cape Town and the two companies at Bloemfontein:—At Fort Hare—One subaltern, three sergeants, and thirteen rank and file; at Leeuwfontein—one subaltern, two sergeants, and thirty rank and file; at Grahamstown—one sergeant and seven rank and file; at Fort Cox were the headquarters of the wing, under the command of Lieut.-Colonel Cooper, including one field officer, one captain, four subalterns, eight sergeants, five drummers, and one hundred and seventy rank and file; while at Fort White were one subaltern, two sergeants, and forty-seven rank and file.

Being thus split into small detachments, all performing garrison duty, the regiment had no opportunity of gaining distinction in the field, as had the 6th, 73rd, 74th, and 91st regiments; consequently, in the blue books of the period we find little or no mention of the services of the regiment in British Kaffraria during the war of 1851. We have already related how Lieutenant Goff was commended for keeping the Kaffirs at bay round Fort White at the close of 1850. On the 9th of January, 1851, Captain Mansergh, of the 6th regiment, was congratulated by the commander-in-chief for his gallant defence of Fort White against a simultaneous attack by four bodies of Kaffirs, Lieutenant Goff and forty-eight men of the 45th forming part of the garrison.

In June Colonel Cooper's garrison at Fort Cox was increased to 294 of his own regiment, 116 of the 91st regiment, 350 of the infantry levies, and 20 of the Cape Mounted Rifles, making a total of 780 men. He was directed to use his force not only as a garrison, but for making " continuous and vigorous patrols to distress and harass the Kaffirs and Hottentots in and around that part of the Amatola range."

In obedience to these instructions Lieut.-Colonel Cooper advanced through the Umagie valley to Fort Hare and returned, capturing a few cattle and exchanging shots with the enemy. Again, between the 20th and 30th of June, he was out patrolling, but nothing notable occurred. Major Kyle, of the 45th, who was in command at Fort Cox, took out a patrol on the 13th of August, consisting of 152 of his own regiment and 12 of the Cape Mounted Rifles, towards Fort Hare. He was attacked by a large body of Kaffirs, engaged them hotly for two hours, and finally drove them off with considerable loss, losing himself one man killed and three wounded. Ensign Walker, of the 45th, who was with him, was much praised by him for his conduct on the occasion.

On the 2nd of August Captain Vialls, of the 45th, left Fort Peddie with 70 men to escort a convoy of waggons, which he handed over to an officer of the 73rd regiment, and proceeded to return *via* Debe's Nek. There he was attacked in the rear by the Kaffirs, and the skirmish was continued for some distance until the enemy was driven off with some loss. At the end of October a detachment of the regiment was again employed, together with a detachment of the 60th Rifles, and some marines, under Colonel Nesbit, in the Waterkloof district, but little was accomplished, though the enemy was driven from all his positions; and a few days later the work had to be done all over again by another column, when Colonel Fordyce, of the 74th regiment, lost his life.

The year at its close left affairs much as they had been at its commencement. A succession of raids had been made against the enemy, many Kaffirs had been killed, and some cattle seized, but no impression had been made, and the Kaffirs were as firm in their strongholds and as audacious as ever. In March, 1852, Sir Harry Smith was superseded, and General Cathcart was appointed in his place, as we have already related, though Lord Grey's despatch ordering his suspension was written on the 14th of January. In the same month news reached the army of the loss of the "Birkenhead" transport, off Simon's Bay, on the 26th of February. There were but six men of the 45th on board, three of whom escaped by swimming, but it has always been a proud remembrance to the regiment to think that it thus had tangible connection with, perhaps, the finest and most magnificent exhibition of steadiness and discipline which any army in the world can boast of. In the same month a column, under Major Kyle, of the 45th, composed chiefly of his own regiment, some Fingoes, and other levies and a six-pounder gun, moved out from Fort Peddie and took up its headquarters at Tamacka, midway between Line Drift and King William's Town, for the purpose of raiding the district. It was so far effective that the chief, Segoli, came to parley, and seemed more than half inclined to surrender. At the request of the Gaika chiefs, to enable them to consult together, an armistice for three days was agreed to, but nothing came of it, and on the 14th Major Kyle continued his operations. About the same time Sir Harry Smith despatched a force from Beaufort composed of the 45th, 43rd, and 73rd regiments, with some artillery and men from the 6th regiment and 60th Rifles, to raid the Waterkloof. This column had a good deal of skirmishing and a full allowance of hard work till about the middle of the month, when it was considered that the district was subdued, and the column was accordingly sent on to raid in the Amatolas. The new governor, Lieut.-General the Hon. George Cathcart, arrived at King William's Town on the 9th of April, and relieved Sir Harry Smith the next day. He at once opened a new and vigorous policy, establishing small stone-works of the nature of Martello towers in place of the larger posts formerly established, which required considerable garrisons, thus leaving much larger bodies of troops available for patrol duties, and to drive the enemy in any direction which was desirable. This policy was steadily carried out, but it was nearly two years before complete success was attained.

In May three companies of the 45th, with Montague's Horse and eighty Fingoes, left King William's Town for the purpose of capturing Line Drift and establishing a fortified post at Tamacka, which they successfully accomplished; and, Lieut.-Colonel Cooper having been superseded in the command of Fort Cox, took over Tamacka Post, from whence, as late as November, he was co-operating with other patrols until the 9th of March, 1853, when Lieut.-Governor Darling reported from Cape Town that the fighting was all over. After peace had been restored the headquarters of the wing, consisting of three companies, remained at Tamacka Post, and included the following officers:—Major and Brevet Lieut.-Colonel Preston, Captains Leach and Grantham, Lieutenant and acting Adjutant Walker, Ensigns Preston, Beamish, and Kingsley, and Assistant-Surgeon Peake. At Fort Murray was a detachment under Lieutenant Lucas and Ensign Perry; at Fort Gray a detachment under Captain Miller and Ensign Douglas; at Fort Pato a detachment under Lieutenant Rowland and Ensign Webber; and at Fort Glamorgan, East London, a party under Lieutenant Hobbs, afterwards relieved by .Ensign Preston.

During the year 1855 the left wing passed some very creditable inspections by Lieut.-General Sir I. Jackson and Colonel Pringle Taylor.

During the year Captain Grantham took his company to King William's Town, where it assisted in the construction of a handsome and spacious hospital for the natives, another monument to the skill and capacity of the regiment.

In February, 1856, the headquarters of the wing moved to Fort Beaufort, and the various detachments of the line of the Buffalo River, being relieved by detachments of the 73rd regiment, rejoined headquarters, but a detachment under Lieutenant Lucas, with Ensign Beamish and Assistant-Surgeon Lithgow, left to relieve the 73rd detachment at Fort Fordyce, on the edge of the Waterkloof. Lieutenant Lucas was subsequently relieved by Lieutenant Rowland, who was soon after relieved by Captain Leach on his return from civil employment in the Amatolas. On the 1st of April one officer and all the available men were ordered from the depôt at Chatham to join the headquarters at the Cape, and in the same month the wing was inspected at Fort Beaufort by Major-General Mitchell, C.B.

From the 7th of May until the 9th of November a road party of one sergeant and twenty men were employed in repairing and reconstructing the road from the Blinkwater Post to Fort Fordyce, which had been much broken by floods during the wet season.

In July the headquarters of the wing moved from Fort Beaufort to Fort Fordyce, being relieved by the 85th regiment, just arrived from Mauritius, and was again broken up into many detachments.

At the end of the year the depôt was moved from Chatham to Colchester. It had received a good many volunteers from the 55th, then in the Isle of Wight, and its appearance on parade was extremely satisfactory to Colonel Jervis on its leaving his command.

The regiment had now been abroad for more than thirteen years, but the hopes of returning home were diminished by fresh alarms in the Colony, which caused the left wing, during 1857, to move repeatedly between Forts Fordyce and Beaufort. Brevet Lieut.-Colonel Preston was appointed to the command of the district, his son, Lieutenant Preston, becoming his district adjutant. There were also many changes during the year between the officers of the left wing and the right wing in Natal, but nothing of any moment occurred. In April a small party of the right wing in Natal was associated with a party of the Cape Mounted Rifles and the Natal Carabineers in suppressing the rising of a native chief on the border of the Colony, and the detachment afterwards occupied Fort Scott, a post on the frontier. Late in the autumn the regiment sustained a sad loss in the sudden death of Lieutenant Arthur Smith of the light company; he was buried in the military cemetery at Pietermaritzburg.

On the 16th of February, 1858, the right wing took part

in a great demonstration at the opening of the Natal Houses
of Parliament, when Lieutenant Blair and Ensign Beamish
acted for the day as A.D.C.'s to H.E. the Governor.    In
the evening there was an official dinner at Government
House, and a ball, which the officers of the wing
attended.    In August the men of the regiment gave a
theatrical performance to raise funds for erecting a monument in the military cemetery to the non-commissioned
officers and men who died in Africa since the arrival of the
regiment in 1843.    In September the citizens of Pietermaritzburg, headed by Chief-Justice Hardinge, gave a
farewell banquet to the men of the regiment, which had
received orders to return home, in the Town Hall.    A most
interesting feature of the entertainment was the attendance,
as waiters, of the men of the Natal Volunteer Carabineers.

Through some mistake the regiment marched out of
Pietermaritzburg before there was any transport ready to
meet it at Port Natal.    A great " marching in " ball was
given to the regiment by the inhabitants of Durban, and
farewells were exchanged; but in the end the wing had to
march back to Fort Napier on the 11th of December, and
remained there until the 16th of April, 1859.

An event of interest, which happened during 1858, was the
promotion of Sergeant-Major F. W. Guernsey to an ensigncy
in the regiment.    He had joined as a recruit from the 43rd
Light Infantry just before the regiment left England for
South Africa, and had been through the Kaffir war of 1846-
1847, and had served in the Orange River Sovereignty from
the end of 1851 to June, 1852.

During the year the depôt removed from Colchester, and,
after a brief stay at Canterbury, became part of the new 5th
depôt battalion at Parkhurst, under the command of Colonel
Jeffreys, C.B.    Early in April, 1859, the 85th regiment
arrived at Pietermaritzburg to relieve the 45th, and a fare-

well banquet was given to the regiment by the officers of the relieving force and the rest of the garrison.

On the 16th of April the regiment finally quitted Pieter-maritzburg, and marched for Port Natal, completing the fifty-two miles in two and a half days, which was looked upon as one of the best short marches ever made. On the 20th of April the regiment embarked on board the "Himalaya," which sailed two days afterwards, and touched at Port-Elizabeth on the 24th to embark the left wing under Lieut.-Colonel Preston. The wings had been separated for sixteen years, and it is said that one of the senior captains of the left wing had never heard the band or seen the colours of the regiment.

Before the left wing left Fort Beaufort to embark for home the following address was presented by about 150 of the principal inhabitants to Lieut.-Colonel Preston:—"We, the undersigned inhabitants of Fort Beaufort, having learned with feelings of sincere regret that the regiment under your command is on the eve of removal from Fort Beaufort, cannot allow the present opportunity to pass without recording the high esteem in which the officers, non-commissioned officers, and men of your gallant regiment are held by the inhabitants of this town. The morality and civility of your regiment we do not believe to be surpassed by any regiment in Her Majesty's service, and on this account we the more deeply regret that you should have been so soon removed from this station, where they are so universally and deservedly respected. With every sentiment of esteem and respect for yourself and your distinguished corps, and praying that the divine blessing may rest upon and accompany you wherever your lot may be cast, we beg to submit ourselves "—etc., etc., etc.

Whilst the "Himalaya" was coaling at Simon's Bay the officers of the corps had an opportunity of meeting and dining with the officers of their sister regiment, the 59th, or 2nd Nottinghamshire regiment, which had recently arrived at Cape Town from China. It is said to have been the first time that the two regiments had ever met; nor is there any record of their meeting again until the year 1882, at Chatham, when both regiments had been deprived of their time-honoured county connection and title. The "Himalaya" sailed from Simon's Bay on the 30th of April, and touched at St. Helena and St. Vincent, arriving at Portsmouth on the 2nd of June. The regiment came home weak in numbers, many of the men having been left behind as settlers in South Africa. The men who returned wore their beards, and their bronzed, hardy, and workmanlike appearance as they marched out of Portsmouth Dockyard won the admiration of all beholders. They were welcomed on their arrival by Major Gordon and most of the officers of the depôt from Parkhurst.

The officers who landed with the regiment were:—Colonel Cooper, in command; Major and Brevet-Lieut.-Colonel Preston, Major Shaw, Captain and Brevet-Major Griffith, Captains Johnstone, Grantham, Griffin, Leach, and Burrows; Lieutenants Gray, Close, Preston, Webber, Beamish, O'Neill, Hayward, Stubbs, and Smith; Ensigns Kershaw and Guernsey, Lieutenant and Adjutant Blair, Paymaster-Captain Blythe, Assistant-Surgeons Cunningham and Bartley, and Bandmaster Signor Girolamo Faccioli.

# CHAPTER XV

THE regiment, on disembarkation at Portsmouth, proceeded
to Preston, where it remained until December of the same
year, when it was moved to Bradford, finding detachments
at Weedon, Burnley, and the Isle of Man.

Its stay, here, however, was not a long one, for in the
succeeding June it was moved to Aldershot, where it was
placed under canvas on Cove Common until the autumn,
when it took up its quarters in the south camp, with one
company detached in the permanent barracks, the whole
battalion subsequently moving to the north camp. In
the musketry returns of the year 1861 the regiment appeared
as the third best shooting regiment of the army, and in the
following year crossed over to Ireland, being quartered in
the Richmond Barracks at Dublin, where, after a short stay,
it proceeded in the following year to the Curragh.

Colonel Cooper, who had commanded the regiment for nine
years, gave up the command this year, and was succeeded by
Colonel William R. Preston.

In spite of the long periods of foreign service which the
regiment had seen during the century, its stay at home was
not destined to be an extended one; for soon after its return
to Beggar's Bush and Ship Street Barracks, Dublin, in 1864,
it received orders to prepare once more for service in India;
and on the 28th of July the headquarters and the right wing,
under Colonel Preston, embarked at Queenstown on the

"Donald M'Kay," being followed four days later by the left wing, under Major Johnstone, on the "Star of India."

The following officers sailed with the regiment:—On the "Donald M'Kay"—Colonel Preston, Captains Hobbs, John Ingle Preston, Beamish, and O'Neill; Lieutenants Tennant, Hooke, Johnson, Dane, Gage, Chambers, and Watling; Ensigns Garnett, Townley, Reeve, and Cartwright; Lieutenant and Adjutant Callwell; and Assistant-Surgeons Wood and Martin. On the "Star of India"—Major Johnstone; Captains Willoughby, Blair, Hayward, and Barlow; Lieutenants Grey, Kyle, and Peterkin; Ensigns Smart and Bridge; Paymaster Nightingale; Quartermaster Guernsey; and Surgeon Speedy.

The headquarter wing reached Bombay on the 1st of November, and were at once landed, taking up their quarters at Colaba. The left wing, which arrived a few days later, proceeded on the 23rd of November to Neemuch.

The headquarter wing was inspected at Colaba, on the 13th April, 1865, by the colonel of the regiment, Sir Hugh Rose, G.C.B., who expressed his unqualified satisfaction at the appearance of the men. On the 7th of November the wing left Colaba and proceeded to Poona by train, the left wing, under Major Johnstone, joining from Neemuch on the 15th of the same month, the whole regiment being quartered in the Ghorpooree Lines. The move to Poona was much appreciated by all ranks; both wings, and especially the one at Neemuch, had suffered during the year severely from cholera, losing a considerable number of non-commissioned officers and men. A handsome monument was subsequently erected to their memory in the churchyard at Colaba.

In December of this year the regiment at last received its much-coveted title of "Sherwood Foresters," which had been applied for before quitting England for the Cape. The

o

title was conferred in consideration of the distinguished services which the regiment had performed, and was conveyed in the following letter:—

"Horse Guards, S.W.,
"12th December, 1866.

"Sir,

"With reference to your letter of the 7th ultimo forwarding an application from the officer commanding the 45th or Nottinghamshire regiment, of which you are the colonel, I am desired by the field-marshal commanding-in-chief to inform you that, on the recommendation of His Royal Highness, Her Majesty has been graciously pleased to approve of the regiment bearing in future the title of 'Sherwood Foresters,' with reference to the traditions of the county of Nottingham and in consideration of the regiment's distinguished services.

"I have the honour to be, Sir,
"Your obedient servant,
(Signed)    "T. TROWBRIDGE, D.A.G.

"Major-General Drought,
"Colonel of the 45th (Nottinghamshire
"Regiment), 'Sherwood Foresters.' "

Early in 1867 the troubles which had been for some years brewing with King Theodore came to a head. In 1855, after years of constant warfare, Theodore overthrew the last of his antagonists, the governor of Tigré, in the battle of Deraskié, and shortly afterwards was crowned king of Ethiopia, with the title of Theodore the Third. As he and his subjects adhered to the Coptic faith, his accession was the signal for the expulsion of the Catholic missionaries, with Father Jacobis, who had to seek a refuge with the rebels in the northern districts. Soon afterwards Theodore united all the forces he could command, marched against the Mohammedan Gallas, ravaged their country, and obtained possession of Magdala; disturbances, however, soon broke out in the conquered provinces, and as a result of these Mr. Plowden, the consul in Abyssinia, was killed while returning to his post at Massowah, while crossing the River Kaka, near Gondar, in 1860, by a rebel force under Garred, a cousin of the king's. Theodore avenged his death severely by the slaughter or mutilation of about two thousand rebels. Captain Cameron, who was appointed Mr. Plowden's successor, arrived at Gondar in July, 1862, and was received by Theodore in person, for whom he brought a present from the Queen of a rifle and a pair of pistols, with all honour and respect. About King Theodore he found a considerable number of Europeans; six German workmen were settled at Gaffat, near Debra Tabor; there were also three missionaries in the country—Rosenthal, Brandies, and Staiger, who were soon afterwards joined by Mr. and Mrs. Stern, as well as a few Frenchmen and others.

In October Captain Cameron was dismissed by the king, who sent him with a letter to the Queen of England; from Adowa he forwarded this letter to Aden, whence it was despatched to England.

After sending this letter Mr. Cameron proceeded to Bogos, and subsequently to Kassala, where he met Mr. Speedy, whom he sent as vice-consul at Massowah. Thence he travelled to Matamma, where he was taken ill, and subsequently returned to Abyssinia, reaching Djenda in August, 1863.

During his absence M. Lejean, who had been appointed French vice-consul at Massowah, visited King Theodore's court. At the time of his arrival the king was on the point of setting out on an incursion into the province of Godjam, which had revolted. M. Lejean accompanied the expedition,

but, soon tiring of the life, demanded his dismissal from
the king.    Theodore refused to see him, upon which he
unwisely forced his way into his presence, for which he
was put in irons for four-and-twenty hours, and subsequently
sent to Gaffat.   In October Mr. Stern, who was returning to
the court, was seized when stopping at the king's camp at
Woggera, and imprisoned; and on the 13th of November
most of the missionaries, including Mrs. Flood, were arrested,
brought into the King's camp, and loaded with chains.
After a mock trial on the 20th of November, the prisoners
were kept in close confinement.    Subsequently Captain
Cameron himself was summoned to the king's camp, and
not allowed to leave.

On the 4th of January, 1864, Captain Cameron and the
Gondar missionaries were summoned into Theodore's
presence, and rudely interrogated as to there being no
answer to the king's letter to Queen Victoria, as a result of
which they were all ordered to be kept as close prisoners.

On the 4th of February Flad, Staiger, Brandies, and
Cornelius were set at liberty, but Captain Cameron and the
others were retained in captivity, as Cameron declined to
undertake that England should not demand satisfaction for
the insult offered to her envoy.    A few days after M. Bardel,
a Frenchman, was added to the number of the captives.

When the news of Consul Cameron's detention arrived in
England it was scarcely believed.   The Foreign Office, however, at once ordered Mr. Hormuzd Rassam, assistant political
resident at Aden, to proceed to King Theodore, with a
letter from the Queen, and also a letter from the Coptic
Patriarch of Alexandria.    On the 23rd of July, 1864, he
arrived at Massowah, and the following day sent off
messengers to King Theodore, with the letter from the
patriarch, and one from himself, announcing that he was the
bearer of the Queen's letter.   In February, 1865, as Theodore

had returned no answer, it was thought Mr. Rassam's embassy would receive more dignity by the addition of a military officer; accordingly Lieutenant Prideaux, of the Bombay Staff Corps, was attached to it.

On the 25th September letters were received at Massowah from the captives, including one from Captain Cameron, urging Mr. Rassam to go up to the king at once, as his declining to do so would prove of the utmost danger to the prisoners. On the 15th of October Mr. Rassam, with Dr. Blanc and Lieutenant Prideaux, left Massowah, and arrived at King Theodore's camp at Damot on the 25th of the following January, delivered the Queen's letter, and were received with all honour by the king, who expressed his intention of releasing the captives.

On the 12th of March the captives from Magdala arrived at Kourata, whither Mr. Rassam and his party had retired, and on the 13th of April the whole party started for the coast.

On arriving at Zagé Mr. Rassam and his companions were arrested with every indignity by King Theodore's soldiers, and the other prisoners who had departed on the Tankal road were also brought back and confined. On the 17th they were all brought before King Theodore, who dictated a letter to Queen Victoria, which Mr. Flad was appointed the bearer of. He left the camp on the 21st April and arrived in London on the 18th of July, 1866.

The other prisoners were sent to Magdala, where they were confined and put in chains, while Theodore's cruelties and atrocities increased tenfold.

But the captives had not been altogether forgotten by their countrymen. Colonel Merewether, the political resident at Aden, urged upon the Foreign Office the fact that force alone would ever cause Theodore to yield up his prisoners. His representations were not without effect, for towards the

middle of April, 1867, the Government began to consider
the possibility of an expedition to Abyssinia. It is probable
that but for Colonel Merewether's persistence the captives
might yet be lingering in chains at Magdala.   In July,
1867, the Secretary of State for India telegraphed to the
Governor of Bombay to inquire, if an expedition were determined upon, how soon the force could be ready to start from
Bombay; and on the last day of July he telegraphed again,
ordering the governor to begin the collection of transport;
and on the 13th of August the expedition was finally decided
upon, and Sir Robert Napier appointed commander-in-chief.
The force he selected for the expedition was composed as
follows:—The 3rd Dragoon Guards, the 1st battalion 4th
regiment, the 26th, 33rd, and 45th regiments, the 3rd Bombay
Cavalry, a regiment of Sind Horse, the 10th and 12th
Bengal Lancers, nine regiments of native infantry, two regiments of Punjab Pioneers, and several companies of sappers
and miners; G battery 14th brigade R.A., under Major
Murray; two batteries of mountain artillery, under Colonel
Milward; a Bombay native mountain battery, a mortar train,
the 5th battery 25th brigade R.A., and a rocket brigade
manned by sailors.

The advanced brigade left Bombay on the 21st of October,
and landed at Annesley Bay on the 30th of the same month,
under Colonel Merewether.   This brigade established itself
at Senafé on the 5th of December, and early in December
another brigade was despatched from Bombay, while on the
2nd of January, 1868, Sir Robert Napier arrived at Annesley
Bay to assume command of the expedition.   The earlier days
of the expedition were spent in making good the road to
Senafé, seven thousand feet above the sea, and providing
water *en route*.   Great difficulties were encountered in
making the road good through the Sooro defile, and excellent
work was done here by a detachment of the Belooch regiment

and the sappers and miners.    By the 25th of January the commander-in-chief considered matters sufficiently pushed forward to justify the commencement of an advance on Antalo; accordingly, orders were sent to hurry on the embarkation of the remainder of the force.

On the 16th of January, 1868, the regiment, which had been armed with the Snider rifle in the previous November, embarked at Bombay on board the ships " Canova," " Gavin Steel," and " America," the following officers sailing with the expedition:—Colonel Parish in command, Colonel Preston having retired during the previous year; Majors Hobbs and Griffin; Captains Close, Hayward, Beamish, and Callwell; Lieutenants Johnson, Lefroy, Bailey, Pollard, Reeve, Lambard, and Curtis; Ensigns Humfrey, Grubb, Bagenal, and Heath; Lieutenant and Adjutant Gage; Quartermaster Guernsey; and Doctors Finnemore, Wood, and Carew.    Captain de Thoren had gone to Abyssinia with the advanced brigade, and was there acting as D.A.Q.M.G.

The following morning the sad news was received of the death of Captain Barlow, who had left Poona a short time before on sick leave.

The headquarters landed at Annesley Bay on the 2nd of February, and went under canvas in the neighbourhood of the landing-place, remaining there until the 4th of March, when four companies, under Major Griffin, moved to a spot about ten miles off ; and on the 9th of March the headquarters and six companies, 501 strong, commenced their march to Antalo.    Of the remaining companies two were ordered to Adigerat, while two remained at the base.

Antalo was reached on the 24th of March, where orders were received to push on to the front as quickly as possible. Lake Ashangi was reached on the 28th, and Dildi on the 31st, where the sad news was received of Lieutenant Bailey's death.    The regiment enjoyed a well-earned rest for the next three days; it had marched 303 miles in twenty-two days, with only one halt, over a practically roadless country, exceedingly mountainous and of considerable difficulty.

Meantime the advanced brigades of the expedition had been pushing on. Sir Robert Napier had arrived at Senafé on the 29th of January, where he had established a secondary base; and on the 18th of February the pioneer force of the army had taken possession of Antalo, by which time King Theodore had retired on Magdala, where he had confined all the prisoners in the fortress.

At Adagabi, two marches beyond Adigerat, the commander-in-chief found it necessary, on account of the weakness of his transport and the consequent difficulty of moving up supplies, to send all the Indian followers and the officers' servants back to Zulla; the baggage being strictly limited to 75 lb. for each officer and 25 lb. for each soldier. In the advance from here great difficulties were encountered in getting the guns along; in many places they had to be unlimbered to allow them to be got round the sharp angles of the track, and the horses taken out and led up singly, while in many places the gunners, assisted by the men of the 10th Native Infantry, had to drag the guns by hand for a considerable distance.

On the 25th of February Sir Robert Napier had an interview with Kassai, Prince of Tigré, who advanced to meet him at the head of a considerable force. A review of the troops present was held to impress the Prince, who parted from the commander-in-chief with expressions of friendship, and, what was more important, undertook to deliver weekly about sixty thousand pounds of wheat and barley.

At Antalo a third depôt was formed, and a stone wall with flanking defences built round the camp to secure the stores. Here all troops of the first division who were not yet up were ordered to push on at once to the front, and all other preparations necessary for an immediate advance were made. The commander-in-chief moved forward from Antalo on the 12th of March, the route lying over one of the great passes of the country, about 10,000 feet in height. On the 15th he assumed personal direction of the pioneer force, and marched with it from Antalo to Makhan, improving the road on the way. On the 20th Ashangi was left, and two days later Lat was reached. Here new arrangements were made for the distribution of the troops. The three brigades of the first division were composed as follows:—

First Brigade—

    3rd Sind Horse.
    A Battery 21st Brigade R.A.
    4th Regiment.
    23rd Punjab Pioneers.
    10th Company R.E.
    Wing of the 27th Native Infantry.

Second Brigade—

    3rd Bombay Light Cavalry.
    B Battery 21st Brigade R.A.
    Naval Rocket Brigade.
    33rd Regiment.

Third Brigade—

    G Battery 14th Brigade R.A.
    3 and 4 Companies Bombay Sappers and Miners.
    Two companies Punjab Pioneers.
    K Company Madras Sappers and Miners.
    One company 33rd Regiment.
    One company 4th Regiment.

The third brigade was ordered to be employed in making the road onward to Magdala practicable for laden elephants, and then to join the other brigades.  The other troops on the way up, including the 45th regiment, were to be passed on to join the first and second brigades.

On the 5th of April the 45th arrived at Santara, only eight days after Sir Robert Napier and the first brigade. Here orders were given to leave all kits behind; the officers were to be quartered ten in a tent, and the men to march with nothing but their greatcoats, a waterproof sheet, and a blanket.  On the 8th the regiment, after a most arduous and trying march of seventeen miles in great heat, reached the summit of the plateaux of Dalanta, where they joined the advanced troops under the commander-in-chief.  Their march had been a memorable one, as ever since leaving Antalo they had practically accomplished a forced march every day.  The *Saturday Review* wrote:—"The march of the 45th regiment will live in the annals of war as a feat of pluck and endurance rarely paralleled"; while Lord Malmesbury, in the House of Lords, subsequently said— "The march of the 45th is one of the most extraordinary on record.  Having been detained in the rear, and being anxious to come as soon as possible to the front, they marched 300 miles in twenty-four days, and accomplished 70 miles in four days over a pass 10,500 feet high."

The officers were most hospitably received and entertained at dinner by the officers of the 4th King's Own.  On the following day the whole force moved forward five miles across the plain to the summit of the descent into the valley of the Bashilo, where it encamped within sight of the heights of Fahla, Selassie, Islamgi, and Magdala, around which the army of King Theodore could be clearly distinguished.

The mountain mass of Magdala forms a crescent, of which Magdala itself is the eastern horn and Fahla the western; midway between the two, in the centre, lies the plateau of Selassie, Magdala and Selassie being connected by the saddle

of Islamgi, and Selassie and Fahla by the saddle of Fahla.
The highest of these plateaux is Magdala, which rises to a
height of over 9000 feet above the sea and 3000 feet above
the ravines of Menchara and Kulkulla.     From the foot of
the Fahla saddle the Wurki-Waha valley runs down to the
Bashilo; up this ravine Theodore had constructed the road
by which he had dragged his guns into position at Fahla.

While the British army had been approaching Magdala
King Theodore had crossed the Bashilo, and on the 25th of
March pitched his camp upon Islamgi, where he remained
to await the attack of the British.

On the 9th of April he issued fresh arms to his men and
then took to drinking, and while intoxicated ordered the
massacre of some three hundred and eighty prisoners, most
of whom were hurled alive over the precipice.

Before daybreak the next morning he assembled his army,
and ordered the road to be prepared for the passage of his
guns from Magdala to Fahla, where he posted four large
guns and four smaller ones.

On the same day the British army moved across the
Bashilo; the 3rd Bombay Light Cavalry, the 3rd Sind Horse,
and the 12th Bengal Cavalry were placed to hold the Bashilo,
while the remainder of the force moved across, the 2nd
brigade being ordered to remain in the bed of the river in
support.

The 45th, after crossing the river, were halted in a
narrow and stony valley, and about 4.50 p.m. were startled
by the sound of heavy firing.  The regiment at once stood
to arms just in rear of the 33rd regiment, when Colonel
Thesiger, the adjutant-general of the first division, rode up
with orders for the regiment to advance as soon as the moon
was up.    The firing had been caused by an attack by
Theodore's forces upon the advanced guard, consisting of
the 4th regiment, the 23rd Punjabees, the naval brigade, and

Penn's mounted battery.    Under cover of the guns his
army had descended in great force to attack, but were
repulsed with a loss of 700 killed and 1200 wounded.  On
our side Captain Roberts, of the 4th, and fifteen men were
wounded.  About 11.30 the same night the 45th, with the
33rd regiment and Murray's battery, commenced the ascent
from the Bashilo, and joined the advanced troops on the
heights before daybreak.  On the morning of the 11th King
Theodore sent Lieutenant Prideaux and Mr. Flad to Sir
Robert Napier, offering to release the captives provided he
was left in peace.  Sir Robert Napier's reply, contained in
a short letter, demanded nothing short of unconditional
surrender.  As a result the King released all the European
captives and sent them to the British camp, but refused to
surrender or to give up his guns.  On the 13th of April,
Easter Sunday, the expeditionary force paraded for the
assault of Magdala, and before evening the fortress was in
our hands with scarcely any loss, and King Theodore and
the remainder of his army utterly destroyed.

The account of the engagement cannot be better described
than in the words of Captain (then Lieutenant) Reeve, of the
regiment, from whose diary the following is taken:—

" At 9.30 the advance sounded; the 33rd went first, then
the 45th and the 10th N.I., these regiments forming the 2nd
brigade, with Twiss' and Penn's mountain batteries carried
on mules, and Murray's Armstrong guns and the rocket
battery and naval brigade; the Beloochees and the 23rd
Punjabees followed.    Winding along the road up Fahla
we expected every moment that Theodore's artillery would
open on us, for it was from this hill that he had fired on us
on Good Friday; but our progress was undeterred by a single
shot, and we got to the top of the hill, on the part called
Selassie, in about three hours, as quietly as if we had been
making an ordinary march.  .  .  .  Had we met with any

opposition on our way we might have suffered great losses, for the position of Islamgi, Selassie, and Fahla was a most formidable one. As soon as I had scrambled to the top of the hill I found all the chiefs giving up their arms to General Merewether, who was assisted by Mr. Flad, one of the captives, and Mr. Waldemar. B Company was then told off to take care of them, under Captain Preston; I was attached to this company for the day, as Goad, my senior, had charge of I Company, which I had hitherto led. The chiefs had surrendered, and were giving up their arms and those of their followers by hundreds. There were spears, matchlocks, a few rifles, knives, daggers, shields, bandoliers. Occasionally they showed some reluctance when compelled to part with a favourite weapon, but otherwise they readily complied with General Merewether's orders, and in some instances appeared quite friendly, offering one a pinch of snuff and taking hold of one's hand, so that it seemed as if the poor creatures were glad to get out of Theodore's hands, and looked on us in a measure as their protectors. . . . All this time the regiment was drawn up about two hundred yards from us; the colours of the 33rd, 4th, and 45th had been planted on the heights. There still remained the great stronghold of Magdala for us to take possession of, in which it was now ascertained that Theodore, resisting to the last, had shut himself up with a few chiefs who remained faithful to him. Captain Hayward's company was now sent to relieve Captain Preston's in charge of the arms, and the latter rejoined the regiment, which was drawn up in a line facing Magdala, of which place we now got our first view. It was about three-quarters of a mile off, and, standing completely by itself, appeared to be a massive lump of rock, and to the naked eye there was no apparent road up to it. Down on the intervening plain we could see the 4th King's Own lying down, and a battery of artillery which soon began to play on the fortress, all the rest of the artillery joining in. The principal object of their aim was the gateway, but the practice was very bad; we all took great interest in the shots, particularly in the rockets.

"After the firing had lasted about an hour the advance sounded, and we marched down the hill, preceded by the 33rd; by this time Hayward's company had rejoined the regiment. The descent was short, but very steep. We passed the 4th regiment, and then the 33rd skirmished, and we formed quarter column in rear of them and advanced rapidly. The 33rd, having kept up a tremendous fire with their Snider rifles against the gate, then commenced the ascent of Magdala in fours, when we deployed into line immediately under the fort, and the left sub-division of Hayward's company (No. 7) opened fire.

"Some shots were fired at us from the gateway, but no damage was done, as all the shots went over our heads. For some little time the gate could not be forced, as it was filled with large stones, and our pioneers were called for. Corporal Cherritt and Private Kirkby responded, and soon forced an entrance. The Madras Sappers, who preceded the 33rd, had forgotten the powder bags. The 33rd cheered loudly as they entered the place, and a signalling flag of the Engineers was waved over the gate. Only four men of the 33rd and Major Prichard were wounded."

[The exact official return of the casualties is given as one officer and nine men wounded, besides three officers and two men who were slightly hurt near the gateway.]

"Soon after the gate was opened Sir R. Napier and staff, and Sir C. Staveley, went up, followed by us, the 4th King's Own, and the naval brigade. About eighty Abyssinians were killed, mostly by shells; about the gate we found ten dead bodies, and behind the second gate I saw the lifeless body of

Theodore, his face much disfigured with blood, his dress the ordinary white gown.

" When we arrived at the summit of Magdala the fighting was all over. The 33rd were busily engaged in turning the natives out of their houses. . . . After piling arms our men went in search of food, and soon brought in quantities of fowl, flour and *tedge* (the beer of the country), which they took from the king's brewery; also honey and a few chupaties. The commander-in-chief then made a circuit of the place, and, after complimenting the 33rd on their gallant behaviour, returned to his camp, leaving only the 33rd and ourselves in Magdala. . . . Twenty-six guns were taken; six were just below Magdala, and had we given Theodore another day he would probably have got them into position."

On the following day the regiment quitted Magdala and marched down the hill to Fort Selassie, where it went into camp.

It has been asserted that the force engaged in the capture of Magdala was unnecessarily large. As events turned out no doubt a smaller force would have sufficed; yet, had Theodore held the summit of his mountain fastness and defended it properly with the weapons at his disposal, the force of the assailants would have been no more than adequate for success. The action of Arogi, too, was anything but a skirmish against ill-disciplined or barbarous assailants. The ground was so broken and covered with tangled brushwood that it was fortunate that the British regiments escaped surprise. The enemy, too, sacrificed his advantage when he descended to fight in the low ground instead of remaining on the hill, and even there he was not repulsed without some skill and valour.

During the next day or two the regiment was employed in protecting the surrendered arms at Selassie, and in providing fatigue parties to bring down prize stores, guns, etc., from Magdala, and on the 17th marched off to the place at which it had halted on Good Friday, the 10th of April. On the same day the fortress of Magdala was burnt and destroyed.

On the following day the British force re-crossed the Bashilo on its homeward route, and encamped that evening on the Dalanta plain, and on the 19th preparations were made for the return of the whole force to Zulla.

On the 20th a review of the expeditionary force was held by the commander-in-chief, after which the loot taken in Magdala was sold by auction and distributed. A handsome silver cross, which now adorns the officers' mess, fell to the lot of the regiment.

At the review Sir R. Napier, in a speech to the troops, said:—" You have traversed under a tropical sun and amidst storms of rain and sleet, 400 miles of mountainous, rugged country. You have crossed ridges of mountains, many steep and precipitous, more than 10,000 feet in altitude, when your supplies could not keep pace with you. In four days you passed the formidable chasm of the Bashilo, and when within the reach of your enemy, though with scanty food, and some of you even for many hours without either food or water, you defeated the army of Theodore which poured down upon you from its lofty fortress in full confidence of victory.

" A host of many thousands have laid down their arms at your feet. You have captured and destroyed upwards of 30 pieces of artillery, many of great weight and efficiency, with ample stores of ammunition. You have stormed the almost inaccessible fort of Magdala, defended by Theodore and a desperate remnant of his chiefs and followers. After you forced the entrance to the fortress, Theodore, who himself never showed mercy, distrusted the offer of it held out to him by me, and died by his own hand.

"You have released not only the British captives but those of other friendly nations.    You have unloosed the chains of more than 90 of the principal chiefs of Abyssinia. Magdala, in which so many victims have been slaughtered, has been committed to the flames, and now remains only a scorched rock. . . .

"The remembrance of your privations will pass away quickly, but your gallant exploits will live in history. . . .

"On my part, as your comrade, I thank you for your devotion to your duty and the good discipline you have maintained throughout.    Not a single complaint has been made against a soldier of fields injured or villagers wilfully molested either in person or property. . . .

"I shall watch over your safety to the moment of your embarkation, and shall, to the end of my life, remember with pride that I have commanded you."

The 2nd brigade, which included the 45th regiment, marched off for the coast on the 21st; Moojah was reached on the 26th, when Major Hobbs was appointed provost-marshal; Ashangi on the 4th of May, and Antalo on the 11th of the same month, where the regiment was played into camp by the band of the 33rd.    The march was resumed on the 14th, and Senafé reached on the 23rd; two days were spent here, and the Queen's birthday celebrated by a review on the 25th.    On the 29th the regiment embarked at Zulla, and, sailing on the 31st, reached Bombay on the 12th of June; here orders were received for it to continue its voyage to Madras, which was reached on the 28th.    On the following day it landed, and was played into barracks by the band of the 60th Rifles.

Thus ended the campaign in Abyssinia, the last occasion on which the regiment, as the 45th regiment, was called upon for active service; and although in this almost bloodless campaign it had little or no opportunity of proving that it

P

could still maintain its old reputation for steadiness under
fire, it had ample opportunities of showing that it could
still surmount those enemies often more to be feared than
the open foe—natural obstacles, hard work, and starvation.
The whole campaign was not the easy walk-over that many
imagine; " the difficulties it entailed would have been more
apparent had their reduction been less skilfully accomplished, and the danger and possibility of disaster would have
been more manifest had they been less skilfully guarded
against."

On arrival in Madras the regiment took over quarters in
Fort St. George, and but little occurred to break the
monotony of the daily routine beyond the visit of the late
Duke of Edinburgh, on board H.M.S. " Galatea," in 1870,
when the customary guards of honour, etc., had to be
provided.

A sad accident happened to Major Hobbs during the stay
at Madras. He was thrown from his horse, invalided home
in consequence, and died shortly after his arrival in England.

In 1871 Colonel Woodbine Parish, C.B., retired from the
command of the regiment, and was succeeded by Major
Griffin.

In January, 1872, the regiment left Madras and embarked
for Burmah; the right wing and headquarters, under Major
Griffin, proceeding up country to Thaetmyo, the left wing
under Major Close, going to Tonghoo. On the arrival of
the right half battalion at Rangoon, the band was detained
there, owing to the visit of the Viceroy of India, the Earl
of Mayo, who was making a tour of inspection, which ended
fatally on the 8th of February in the Andaman Islands,
where he was assassinated by a convict.

During its stay up country in Burmah the regiment
suffered a good deal from the ravages of disease, especially
the wing at Tonghoo, which lost a considerable number of

non-commissioned officers and men, and three officers—
Lieutenants Kyle, Smith, and Montgomery; while Colonel
Griffin, who had been invalided, died on board a P. and O.
steamer between Calcutta and Madras on the 10th of
September, 1873, and was buried at sea.

In January, 1874, both wings were once more re-united at
Rangoon, and in February Colonel Mark Walker, V.C., of
" the Buffs," who had been appointed to the lieut.-colonelcy
vacant by the death of Colonel Griffin, assumed the command
of the regiment.

Colonel Walker, who originally joined the 30th regiment,
served with great distinction in the Crimea, and was awarded
the Victoria cross for gallantry at the battle of Inkerman,
where, to encourage his men, he jumped over a stone wall
in the face of two Russian battalions, and thus led on his
regiment and repulsed the enemy.

In February, 1875, the regiment quitted Burmah, landing
once more in Madras, and proceeded by rail to Bangalore,
a change much appreciated by all ranks.

In August of this year Major John Ingle Preston succeeded to the command of the regiment, in place of Colonel
Walker, V.C., who was appointed to a brigade command.
It is rather a remarkable coincidence that he should have
succeeded his father in the command after the comparatively
short interval of eight years.  In December Colonel Preston
was invalided home, and Major Adams took over the command; while a year later Major Hayward, who had been at
home, rejoined and took over command from Major Adams.

In the autumn a sad event occurred which cast a gloom
over all ranks.  Private Fuller, of A Company, was murdered in the barrack room by a comrade named Neil, who
shot him dead while he was sleeping on his bed.  The
motive of the murder was jealousy, and Neil was more or
less under the influence of drink when he committed the

crime. As the battalion was stationed beyond the jurisdiction of the High Court, Neil was tried by general court-martial, which sentenced him to death. The sentence was carried out at daybreak, in front of the whole regiment and detachments from the other corps stationed in Bangalore, and made a deep impression upon all who witnessed it.

During the year a detachment under Captain M'Cleverty was sent to Trichinopoly.

On the 1st of January, 1877, the regiment, in common with the entire garrison, paraded at mid-day to attend the proclamation of Her Majesty Queen Victoria as Empress of India. After the reading of the proclamation by Mr. Gordon, the acting commissioner of Mysore, a salute of one hundred and one guns was fired by the artillery, and a *feu-de-joie* by the infantry, followed by three cheers for Her Majesty. A medal was struck commemorative of the event, and one distributed to each corps and battery serving in India, the one which fell to the lot of the regiment being presented to Sergeant-Major Griffiths, who subsequently became quartermaster of the 4th battalion, Oxfordshire Light Infantry.

The famine which devastated the Madras Presidency was especially bad in the neighbourhood of Bangalore throughout the summer of 1877, and the cholera, which followed in its wake, caused fearful havoc among the native population; on this occasion, however, good fortune favoured the regiment, not a single case occurring in its ranks.

In the autumn the welcome news was received that the regiment was for home in the early part of the following year, and for a time volunteering for other battalions in India, and the arrangements for the transfer to the linked battalion of the few short-service men in the ranks, were the only breaks in the ordinary monotony of garrison existence in India.

Early in 1878 the depôt moved to Shorncliffe to await the

arrival of the service companies from India. Under Mr. Cardwell's scheme of linked battalions, which came into operation in 1873, the regiment had been linked with the 17th regiment—this being the only link which comprised three battalions—and had followed the fortunes of that regiment since that date, moving successively from Woolwich to Aldershot, the Curragh, Athlone, and Newport in Monmouthshire. Both battalions of the 17th regiment being abroad at this date, the two depôts accompanied the depôt of the regiment to Shorncliffe. On the 4th of February, 1878, the left half battalion, under Major Adams, left Bangalore for home, being followed on the following morning by the headquarters and right half battalion, under Major Hayward. Both half battalions were played to the station by the bands of the native regiments in garrison and the 14th Hussars; the latter regiment generously supplying refreshments for the men and breakfast for the officers on the platform at the railway station.

The regiment halted for three days at Poona, and leaving there in two parties on the 13th of February, embarked at Bombay in H.M.S. "Euphrates," and sailed for home the following morning.

The voyage was to some extent an eventful one. The war which had been going on between Russia and Turkey had reached about this time an acute stage, and threatened to involve other nations, particularly England, in the struggle. On arrival at Malta affairs looked so threatening that the regiment was detained there for upwards of a week, and every one was on the tip-toe of expectation, hoping that their destination might be east rather than west; but the atmosphere cleared for a while, and the voyage home was continued. England was not, however, to be reached without further delay; heavy weather was encountered just to the northward of Cape Finisterre, and the "Euphrates" was put about and ran for Vigo, which she entered on the 17th of March. After two days' detention here the voyage was resumed, and without further adventure Portsmouth was reached on the 22nd of the same month, the regiment landing and proceeding to Shorncliffe on the following morning.

During the voyage the reputation possessed by the regiment, which has more than once been previously mentioned, for settling down at sea was more than sustained, and strongly testified to by the captain and officers of the "Euphrates," who were loud in their praises of the behaviour, discipline, and efficiency displayed by the regiment during the time it was on board; while the good feeling prevailing between the regiment and the crew was evidenced by the latter turning out to give hearty cheers for the regiment as the train steamed away from Portsmouth dockyard.

The following officers landed with the battalion:—Major Hayward in command, Major Adams; Captains Dillon, Martin, Baines, Gage, Lambard, Wright, and M'Cleverty; Lieutenants Fort, Dowse, Neild, Coney, Bulpett, Pearse, and Todd-Thornton; Sub-Lieutenants Dalbiac and Swann; Lieutenant and Adjutant Jones, Quartermaster Higgins, and Surgeon Bushe.

Upon arrival at Shorncliffe Lieut.-Colonel Preston resumed command, and two days later the regiment was inspected in a deep snow by Major-General Cameron, who addressed it at some length, dwelling especially upon the subject of insubordination, a crime practically unknown in the battalion.

Affairs in the East again became critical about this period, and the reserves were called up throughout the country, with the result that some 600 militia reservists, from the Leicester and Nottingham regiments of militia, joined the regiment, remaining for upwards of four months, until after the Conference of Berlin, when peace was proclaimed between Russia and Turkey, and affairs settled down.

This naturally made the regiment exceedingly strong, and at the searching inspection by Major-General Cameron, during the summer, there were nearly 1400 men on parade.

The loss of the German warship "Grosser Kürfurst," by collision about two miles off Sandgate, in the early summer, in which a very large number of lives were lost, provided a good deal of work for a time in the furnishing of funeral parties at the interments of the bodies which were washed ashore.

The disaster at Isandlana in January, 1879, necessitated the rapid despatch of a force to subdue the Zulus; and the regiment was called upon to provide eighty volunteers for service with the 58th regiment, then lying at Dover, which had been included in the expeditionary force. The men provided were a fine lot of old soldiers, and elicited much favourable comment upon their splendid physique on their arrival in Dover. Many of them fell, two years later, at Laing's Nek and Majuba Hill; and one man, Private Godfrey, was awarded the medal for distinguished conduct in the field, during the former action, for his gallantry in endeavouring to assist Major Hingeston, of the 58th, who was mortally wounded.

During the spring the regiment was moved to Aldershot, and quartered in the South Camp, where it remained until the autumn of 1880, when it was moved down into the West Infantry Barracks.

In August Colonel Preston relinquished the command of the regiment, after completing his period of five years in the appointment, and retired from the army with the rank of major-general. He was succeeded in the command by Major Hayward.

Our history now draws to a close.   Early in 1881 Mr.
Childers introduced his scheme of army organisation,
apparently with the object of killing every cherished tradition
of the service and obliterating everything that could lead the
men of to-day to remember the glorious deeds of those who
had gone before them.   Numbers were abolished throughout
the army, most regiments were deprived of their distinctive
facings, and territorial titles were re-assorted on a method
so incongruous as to pass human understanding.

The regiment suffered perhaps more severely than any
other; from being the 45th, Nottinghamshire regiment, it
became the 1st Battalion Derbyshire regiment, while its
time-honoured facings of Lincoln green were replaced by the
white facings made common to all English regiments.

These changes came into operation on the 1st of July; on
the previous evening, after mess, the colours were brought
out and carried in triumph round the barrack square; the
Queen's colour being carried by Major Callwell and the regimental colour by the adjutant, Lieutenant Jones.   The men
turned out of their barrack rooms, and, after cheering
vociferously for their moribund number and the old facings
of the Lincoln green, joined as one man in singing "Auld
Lang Syne" in token of that feeling which, though it may
appear a slight thing to the outside world, is more deeply
cherished than words can express by the members of a
regiment—a pride in the honourable traditions of their
corps.

The next morning the 45th regiment had ceased to exist,
the ". 45" on the men's shoulder-straps had given place to the
word "Derby," and men looked unfamiliarly at the white
collars and cuffs of their comrades' tunics.

Thus, by the stroke of a Minister's pen, the colour which
nearly one hundred and fifty years before had earned for

the regiment the sobriquet of the "Green Marines," and the county appellation under which, for upwards of a century, without the occurrence of a single incident to mar its good name, the regiment had maintained the best traditions of the British army and the honour of England in four continents, were swept away as if unworthy of a moment's consideration.

## APPENDIX A.

THE first official list obtainable of the officers of the regiment is the one published in 1754, a copy of which appears below, as well as the last list, from the Army List of June, 1881, before the regiment lost its distinctive number and county designation, under Mr. Childers' Army Scheme of that year.

### 1754.

| | |
|---|---|
| *Colonel* | Hugh Warburton. |
| *Lieut.-Colonel* | Montagu Wilmot. |
| *Major* | Hungerford Luttrell. |
| *Captains* | Alexander Murray. |
| | William Cottrell. |
| | William Walters. |
| | Patrick Sutherland. |
| | William Powell. |
| | George Croxton. |
| | James Clarke. |
| *Captain-Lieutenant* | Ralph Hill. |
| *Lieutenants* | Thomas Vaughan. |
| | James Ouchterlony. |
| | John Mitchell. |
| | James Cunningham. |
| | Gilfred Collingwood. |
| | John Pinhorn. |
| | Broderick French. |
| | Henry Dugdale. |
| | Richard Stevens. |
| | Winkworth Tongue. |

| *Ensigns* | Richard Bulkley. |
|---|---|
| | William Needham. |
| | Charles Husband Collins. |
| | John Bevan. |
| | Walter Crosbie. |
| | John Bowen. |
| | John Wright. |
| | Edward Leigh. |
| | William Hall. |
| *Chaplain* | Robert Brereton. |
| *Adjutant* | William Powell. |
| *Quartermaster* | Herbert Laurence. |
| *Surgeon* | Richard Veale. |

### 30th June, 1881.

| *Colonel* | Lysons, Sir D., K.C.B., Gen., | 25th Aug., '78 |
|---|---|---|
| *Lieut.-Colonel* | Hayward, Hen. B., | 4th Aug., '80 |
| *Majors* | Adams, Thos. E., | 4th Aug., '75 |
| | Callwell, Rob. J., | 4th Aug., '80 |
| *Captains* | Hooke, Henry Hodson, | 23rd Oct., '67 |
| | Dillon, Rob. Hen. M., | 16th May, '65 |
| | Wright, Alexander, | 1st April, '75 |
| | M'Cleverty, James, | 4th Aug., '75 |
| | Cobbold, Geo. Hen. | 19th Dec., '77 |
| | Thrupp, Morton. F. S., | 29th June, '78 |
| | Fort, Geo. H., | 30th Jan., '80 |
| | Lloyd, Arthur Clifford, | 15th Aug., '80 |
| | Hume, John W. Thring, | 18th Dec., '80 |
| | Carter, Edward Aug., | 19th Feb., '81 |
| | Dowse, Edward Cecil, | 23rd Mar., '81 |
| | Hudson, Herbert E., | 4th May, '81 |

| *Lieutenants* | Jones, Geo. Alfred, Adj., | 23rd May, '74 |
|---|---|---|
| | Neild, Hen. Ed., | 11th Feb., '75 |
| | Coney, Wm. Bicknell, | 11th Feb., '75 |
| | Dalbiac, Philip Hugh, | 11th Feb., '75 |
| | Swann, John Chris. (prob.), | 10th Sept., '75 |
| | Bulpett, Art. Dolben, I. of M., | 20th Nov., '75 |
| | Littledale, Hen. Art., | 12th Nov., '73 |
| | Todd-Thornton, J. H. B., | 17th Jan., '77 |
| | Pearse, Napier L., | 24th Mar., '77 |
| | Watts, Charles N., | 1st Jan., '81 |
| *Second-Lieutenants* | Reeks, James A., | 11th Aug., '80 |
| | Kilbee-Stuart, Ronald R., | 22nd Jan., '81 |
| *Paymaster* | Stuart, T. E., Hon. Capt., | |
| *Instructor of Musketry* | Bulpett, A. D., | |
| *Adjutant* | Jones, G. A., | |
| *Quartermaster* | Higgins, Maurice, | |

## APPENDIX B.

The following is the list of colonels of the regiment from 1745 to 1881:—

1745. Hugh Warburton, lieut.-general.
1762. Hon. Hugh Boscawen, major-general.
1763. Hon. John Boscawen, major-general.
1768. William Haviland, major-general.
       (Became lieut.-general in 1772.)
1785. Sir John Wrottesley, Bart., major-general.
1788. James Cunningham, major-general.
1789. James Adeane, major-general.
1803. Cavendish Lister, lieut.-general.
       (Became general in 1804.)
1824. Earl of Cavan, K.C., general.
1838. Sir William Henry Pringle, G.C.B., general; died 30th December, 1840.

1841. Sir Fitzroy Jeffries Grafton Maclean, Bart., general.

1848. Sir Colin Halkett, K.C.G., G.C.H., general.

1857. Thomas Brabazon Aylmer, general.

1859. Sir Hugh Rose, G.C.B., major-general.
 (Became lieut.-general in 1863.)

1866. Thomas Armstrong Drought, major-general.

1869. Sir Frederick Horn, K.C.B., major-general.
 (Became lieut.-general in 1870.)

1876. Henry Cooper, lieut.-general.

1879. Sir Daniel Lysons, K.C.B., lieut.-general.


## APPENDIX C.

List of lieut.-colonels commanding from 1739 to 1881 :—

1739. Colonel Robinson ; killed at Carthagena, 1741.

1741. Lieut.-Colonel Frazer.

1741. Colonel Houghton ; died in 1748, when governor of Pendennis Castle.

1745. Colonel Hugh Warburton.

1755. Colonel Montagu Wilmot.

1762. Lieut.-Colonel John Tullikens.

1772.  ,,  Hon. Henry Monckton.

1780.  ,,  William Gardiner.

1782.  ,,  Hon. Henry Phipps.

1784.  ,,  Francis Dundas.

1788.  ,,  Oliver Nicolls.

1795.  ,,  Bryan Blundell.

 ,,  Alexander Fraser.

1796.  ,,  Bryan Blundell.

 ,,  James Montgomery.

1800.  ,,  James Montgomery.

 ,,  William Guard. Joined as a volunteer at Boston; served at the battles of Brooklyn and Whiteplains; promoted colonel in 1806 ; severely wounded at Talavera.

1805. Lieut.-Colonel William Guard.
           „      Lord Frederick Bentinck.
1806.     „      William Guard.
           „      Hon. John Meade.
1812. Colonel William Guard.
    Lieut.-Colonel Hon. John Meade.
           „      Henry Ridewood; killed at Vittoria.
1814. Colonel William Guard.
    Lieut.-Colonel Hon. John Meade.
           „      Thomas Forbes; severely wounded at Salamanca; killed at Toulouse.
1815.     „      Leonard Greenwell, C.B.; severely wounded at Buenos Ayres, Salamanca, and Orthés, and received the C.B. for Salamanca.
1826.     „      Leonard Greenwell, C.B.
           „      Hugh Stackpoole.
1828.     „      Charles A. Vigoreux.
           „      Thomas Shaw.
1834.     „      Charles A. Vigoreux.
           „      Edward French Boys.
1840.     „      Edward French Boys.
1847.     „      Edward French Boys.
           „      Archibald Erskine.
1849.     „      Edward French Boys.
           „      Henry Cooper.
1854.     „      Henry Cooper.
1863.     „      William R. Preston.
1868.     „      Henry Woodbine Parish, C.B.
1872.     „      Charles L. Griffin; died 10th Sept., 1873.
1874. Colonel Mark Walker, V.C.
1875. Lieut.-Colonel John Ingle Preston; retired as major-general in 1880.
1880.     „      Henry Blakeney Hayward; retired as major-general in 1885.

## APPENDIX D.

### THE PENINSULA MEDAL.

As the result of a petition presented to the House of Lords by the Duke of Richmond in 1845, and supported in a brilliant speech by the Duke of Wellington, a general order was issued, two years later, in 1847, by command of Her Majesty, to the effect that a medal should be struck as a reward for the services of the fleet and army in wars between 1793 and 1814, and that applications for the same should be sent in by all ranks who had served between these periods.

The Peninsula medal had on the obverse side the head of the Queen crowned, around are the words "Victoria Regina," and the date 1848 underneath the bust. On the reverse, the Queen in the act of crowning with laurel the Duke of Wellington, with the legend "to the British army, 1793-1814." Clasps for the various actions were attached to the ribbon.

There were numerous claimants for the medal, but only six survivors applied for fifteen clasps, only two of whom could make good their claim, viz., Private James Talbot, 45th regiment, and Private Daniel Loockstadt, 60th Rifles, formerly of the German Legion.

The clasps were for—Roleia, Vimiera, Corunna, Talavera, Busaco, Fuentes d'Onoro, Ciudad Rodrigo, Badajoz, Salamanca, Vittoria, Pyrenees, Nivelle, Nive, Orthéz, and Toulouse, all of which the 45th regiment bear on their colours except Corunna and Nive.

The following officers who had served with the 45th regiment in the Peninsula war received the medal in 1847 :—

Captain T. P. Costley. L.p., 45th regiment. Clasps for Busaco, Fuentes d'Onoro, Ciudad Rodrigo, Badajoz, Salamanca, Vittoria, and Pyrenees.

Bt. Lieut.-Colonel H. A. Fraser (at Waterloo also). Clasps for Corunna, Busaco, Fuentes d'Onoro, Ciudad Rodrigo, Badajoz, and Toulouse.

Lieutenant A. Lowry. H.p., 45th regiment. Clasp for Orthéz.

Major A. Martin, 45th regiment. Clasps for Roleia, Vimiera, Talavera, Ciudad Rodrigo, Nivelle, Nive, and Orthéz.

Lieutenant Chas. Munro. H.p. Clasps for Ciudad Rodrigo, Badajoz, Salamanca, Orthéz, and Toulouse.

Captain James Reid. Clasps for Nivelle, Nive, and Orthéz.

Captain J. H. Reynell. Clasps for Busaco, Fuentes d'Onoro, Ciudad Rodrigo, and Badajoz.

Surgeon W. Smyth, 45th regiment. Clasps for Busaco, Albuera, Pyrenees, Nivelle, Nive, Orthéz, and Toulouse.

Colonel W. Guard. Clasps for Roleia and Vimiera; he also received the special decoration of the Peninsula gold medal.

Major-General T. Lightfoot. Clasps for Roleia, Vimiera, Talavera de la Regna, Busaco, Ciudad Rodrigo, Badajoz, Salamanca, Nivelle, Nive, Orthéz, and Fuentes d'Onoro.

Captain B. G. Humphrey. H.p., 56th regiment. Clasps for Ciudad Rodrigo, Badajoz, Salamanca, Vittoria, Pyrenees, Nivelle, Nive, Orthéz, and Toulouse.

Captain W. Hardwick. H.p., 2nd regiment. Clasps for Fuentes d'Onoro, Pyrenees, Nivelle, Nive, Orthéz, and Toulouse.

Bt. Major J. Macpherson. Clasps for Roleia, Vimiera, Corunna, and Salamanca.

Lieutenant James John Rowe. H.p., 7th regiment. Clasps for Vittoria, Pyrenees, and Nive.

Major James Campbell, 50th regiment. Clasps for Roleia, Vimiera, Talavera, Busaco, Fuentes d'Onoro, Ciudad Rodrigo, Badajoz, Salamanca, Vittoria, Pyrenees, Nivelle, Nive, Orthéz, and Toulouse.

Lieut.-Colonel Chas. Barnwell, C.B. (Adj. 45th regiment), 9th regiment. Clasps for Roleia, Vimiera, Busaco, Fuentes d'Onoro, Ciudad Rodrigo, Badajoz, Salamanca, Vittoria, Pyrenees, Nivelle, Nive, Orthéz, and Toulouse.

Bt. Major F. Andrew, 52nd Light Infantry. Clasps for Roleia, Vimiera, Talavera, Ciudad Rodrigo, Badajoz, and Salamanca.

Captain Richard Colley (Adj., 45th regiment), 1st regiment. Clasps for Roleia, Vimiera, Talavera, Busaco, and Fuentes d'Onoro.

Staff-Surgeon T. Hoggie. Clasps for Roleia, Vimiera, Talavera, and Fuentes d'Onoro.

Captain James Bishop, 23rd regiment. Clasps for Vittoria, Pyrenees, and Nivelle.

Captain Joseph Douglas. Clasps for Nive, Nivelle, Orthéz, and Toulouse.

Lieutenant Thomas Atkins, 73rd regiment. Clasps for Roleia, Vimiera, and Corunna.

From the number of officers who received the clasp for Nive it would seem probable that the regiment, or at least some portion of it, was in action at Nive, though the name is not borne amongst the honours on the colours.

## APPENDIX E.

Lieut.-Colonel Oliver Nicholls, who commanded the regiment from 1788 to 1795, was presented by the Corporation of Wexford with a gold medal for services rendered in 1793 to the town.

On the obverse are the arms and motto of the town of Wexford. On the reverse the following inscription:—" Dedicated to Oliver Nicholls, Lieut.-Colonel of the 45th regiment, by the inhabitants of Wexford, armed to protect the peace thereof, against the insults of an armed mob which appeared before its walls, 11th July, 1793, as a testimonial of their gratitude justly due to his distinguished aid and active zeal in co-operating with them.".

## APPENDIX F.

It was a custom sometimes, before the introduction of a general distribution of war medals by the Government, for the officers of a regiment to have a medal struck, commemorative of a campaign, for distribution among the men of the regiment. There are many such medals in existence, and among others one issued by the officers of the 45th after the Peninsula campaign.

It is in the form of a Maltese cross, with the number of the regiment in the centre, surmounting which are the names of twelve Peninsula victories. The reverse is plain.

# INDEX

Milton Keynes UK
Ingram Content Group UK Ltd.
UKHW022206281223
435087UK00004B/50